Tinkering Puff

The Pop-Up

Origami

Rover

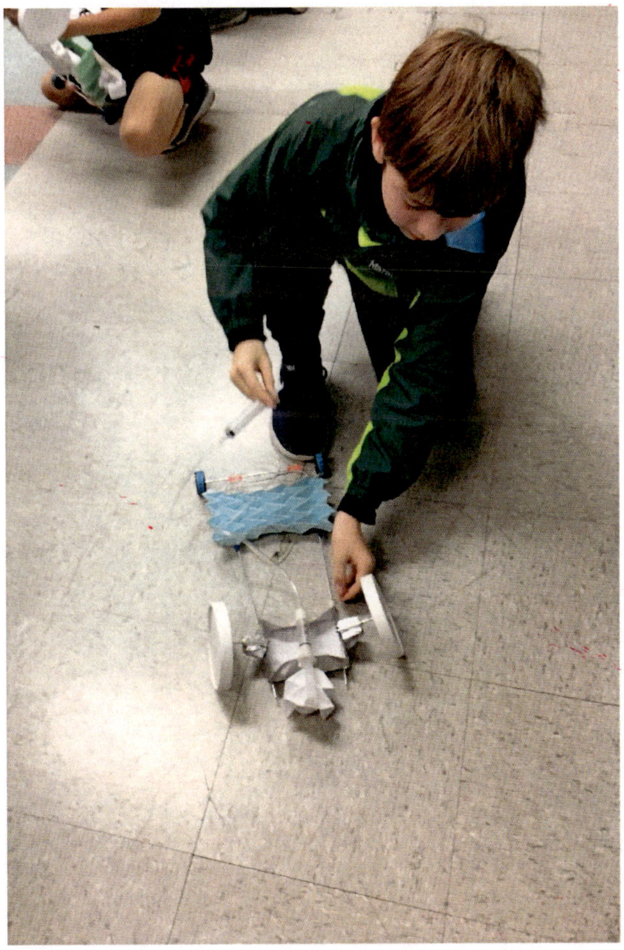

Remembering Children's Contraptions

Meredith Olson, Ph.D.

www.DocOsBooks.com

www.DocOsBooks.com

Printed in USA

Published by:
SAN 299-2701
Glenhaven Publications
4262 NE 125 Street
Seattle, Washington 98125

ISBN 978-0-9984627-8-3 (paperback)

Table of Contents

Preface: Educational Philosophy

For forty years our homegrown engineering event has provided students with experiences in tinkering with physical science, engineering, and artistic design. Most of the ideas advocated by currently popular programs such as STEM, STEAM, and PBL are naturally incorporated because we focus on providing students personal experience with the process of science. We celebrate our clever contraptions in our yearly community event.

It has been said that the typical classroom lesson consists of a teacher lecture followed by independent quiet seatwork. Not here! We prefer active, enthusiastic, focused engagement. From every student.

Our lecture typically consists of one sentence or so. That is followed by students reporting "class data" wherein each and every two-person lab reports their data from the previous day. Every child speaks. Every child listens with interest to the report of others because that informs their assessment of their own data. There are no answer books. The only way to evaluate their own work is by comparing their work to others. Every child writes a "class data" table as the numbers come in so they are able to compare and contrast the information. There are sighs and "yea's" and laughter as the data are reported.

Students are learning cooperatively through carefully engineered student interaction which encourages sharing and engagement by all. No student is left behind. No student is left out. This fundamental scientific method of data sharing creates a dramatic positive difference in the learning and enthusiasm of every child. They can not determine the useful trend in the data without working together. Collaborating. Using gentle trends in mathematics to make meaning of their lab work. Measuring distance, volume, force. A few simple calculations. Not asking for mathematics beyond their comfort level, but naturally integrating numbers in the way we understand the world. Laughing at numbers. Engaging with numbers. Using trends in numbers in our daily explanations. There is an overarching flavor to the classroom that the process of science is enormous fun. Focused, engaging hands-on memorable play. Student notes show how hard they each work and how deeply they each think as they become confident, playful scientists. Alumnae return to say that this method lasts. They have become life-long learners.

I do not give grades in science. Children achieve different levels of excellence and they know it. Some contraptions are ingenious while others pretty much reflect the basic design criteria. This is equality of opportunity, not equality of outcomes. Classroom rules establish the behavioral protocols that allow everyone the freedom to engage in a project and be as clever and productive as they can. It promotes the recognition of individual intellectual property while allowing others to copy and build upon those ideas. ("I am going to use the 'Ben' wheel system.") All students are subject to and accountable to classroom rules that are fairly applied and enforced. Motivation comes from self-fulfillment, appreciation from peers, and general cohort enthusiasm for the entire project.

It is necessary to strike a balance appropriate for the age and maturity of the child, between regulated behavior and freedom of expression. We offer design criteria and specify which materials may be used in a project. Within that framework, when criteria are met (e.g. The rig must move on its own) then the student is free to invent new attributes. The work is entirely done at school. This avoids participation by overzealous parents and friends. If the home wants to be involved they are encouraged to make "home-built" rigs to enter in a separate heat in the final event. The materials have all been provided by the school so there is equality of opportunity. But their zeal and ingenuity provide their differentiated level of outcomes. We intend to create a society in the classroom where each individual is intensely struggling to develop his or her own best rig. We require that the distance rigs move have a witness. The classroom rings with shouts of, "Clear the course!" "Witness, witness!" The enthusiastic self-interest of each child motivates others to produce and improve their own rigs. At the end of the event, each child takes home their own rig. It is theirs to keep. With pride.

It is quite odd to read recommended lessons teaching rubber band cars on the internet. The Science and Health Education Partnership (SEP) of the University of California, San Francisco directs teachers to start by *"telling students that today you will explore what makes things move." Next, they roll a ball and show a battery-powered toy. They show them a wind-up toy and ask them to record observations.* (A quiet Language Arts activity) Students draw it and are asked how the behavior changes if they wind it more. Then they discuss their observations. The lesson concludes asking students to *record key vocabulary on their worksheet and then use the terms to describe how the wind-up toy works.* So far the lesson asks primarily for verbal or descriptive reporting. Classes would be expected to be quiet, orderly and in written response mode.

The next day wind-up wooden cars are introduced. The teacher is asked to, *"Demonstrate how to attach and wind-up the rubber band. Let go of the cart and watch it drive. Have students talk and turn to each other to explain how the car moves, using the science terms from the last lesson. Hear some students report out their ideas."*

The teacher is told to next, *"Ask student how we might be able to make the car go further. Record ideas on board as students share out. Possible ideas might include: winding up the rubber band more times, using different rubber bands, using more than one rubber band. As a class, develop a hypothesis for each of their ideas, write: "The greater the number of times the rubber band is wound up, the farther the car will go." "The greater the number of rubber bands used, the farther the car will go."*

Then, "Show students the hand-out and explain that they will first establish a base line, winding up the rubber band 3 times. Tell them that they will do that 3 times/trials.

Then students will increase the number of times they wind-up the rubber band and *record those results, with 3 trials each time. Students calculate medians. Students write # of turns and median distance on sticky dots and place it on the class graph on board. Discuss the results as class. Ask students to explain their observations/data. Can we predict how far the toy would go with 12 turns?"*

Then, "Read 'Energy makes things happen' book by Kimberly Brubaker Bradley. As you read the book, record the different kinds of energy that are mentioned."

This lesson meets NGSS Disciplinary Core Ideas and NGSS Performance Expectations. And it meets NGSS Science and Engineering Practices: Developing and Using Models , Planning and Carrying Out Investigations , Analyzing and Interpreting Data , Using Mathematics and Computational Thinking , Constructing Explanations and Designing Solutions , Engaging in Argument from Evidence , Obtaining, Evaluating, and Communicating Information

And NGSS Crosscutting Concepts:
Patterns ,Cause and Effect ,Systems and System Models ,Energy and Matter Structure and Function ,Stability and Change

And, it seems to me, it turns a potentially exciting lesson idea into a boring, quiet class activity that is perfunctory and has lost a personal zest for doing the investigation. What ought to be delightful play with toys is turned into a stifling Language Arts assignment. That isn't necessary. Although we cover the same material, that isn't the way we do things here. And that glassy-eyed compliant student response isn't what we get.

First of all, my students each build their own rig. They copy each other as they need to in order to figure out technical difficulties (such as friction on the axle and glue in the bearings) to make it move. When the rig moves, we loudly celebrate! I climb on a step-stool and put the student name on the board, writing how far it went. As others get their rig moving, their name goes on the board. When their distance is improved, a second number of floor tiles is listed after the first. "Yea for you!" The next day class data (listing each student and how far their rig moved) is collected by each student in their journal as I make a matching table on the overhead projector.

It is interesting to me that having students each write the data in their individual journals produces more enthusiasm, more ownership of the data, more referring back to it on future days than having stickies put on a graph on the wall. Thus, my students rehearse their writing skills (as much as the paragraph writers do) but the general level of enthusiasm is much greater. It is magic!

Contributing data the way we do gives every student a voice. They speak aloud and the class listens intently – trying to write down what they say. It allows me to say, "Oh really?" It communicates that I publically listen to every student every day. Individuals matter. They matter individually. If they are absent; if they don't get data; we notice. We need them. Each one. They are not faceless and left behind. Kids don't want to miss science class.

You would think that putting stickies on the wall would be fun. But it isn't. You would think that putting stickies on the wall would produce collegiality and cooperation. But it doesn't. It results in disengagement. It gives away their unique data point and gives it to a personalityless wall display that doesn't care. In my experience, putting stickies on the wall degrades personal enthusiasm. Psychologically, it is like giving away your toy. You are through playing with it personally. On the other hand, each person writing down data as each lab reports, sends a message that they are expected to do their own analysis. Students enthusiastically wait for the report from the next lab and joyfully yell out trends they notice in the data. Class data becomes "their" data. Trends become their own personal discovery (even with the teacher modeling the behavior at the overhead projector). The enthusiasm and levity associated with each contribution to class data is palpable. Science class is personal. Science class is fun.

Instead of making shared data communal, we strive to make shared data personal. Personal ownership. Personal responsibility for novel ideas. A classroom of competent individuals sharing together. Collegial yet individual. The psychological tone of the classroom shifts from "group think" to individual analysis. The difference is subtle but it matters. It makes the difference between finding science boring and finding it energizing fun. And it allows for individual brilliance. A Nobel Laureate has all the data everyone else has, but they reserve for themselves the ownership of their unique analysis of it.

We discuss uncontrolled variables and technical difficulties and offer suggestions to each other. After about eight minutes of class discussion and data review, students scurry to their labs to try to improve their personal contraption. Individual ownership motivates! The room is loud and enthusiastic. Students yell, "Clear the course!" "Witness, witness!" This seeming unruly noise and playful competition produces a far greater love of science class than politely and quietly waiting in line for a turn to use a racetrack for a vehicle loaned to you by the teacher.

I do not require written paragraphs about what they learned. We can communicate in more than text form. In fact, for some kinds of information text is not a useful way. Geometry textbooks before the early 1900's, with text and very few drawings, are painfully difficult to read. For numerical data, data tables and graphs are far better at presenting information, seeing patterns and promoting discussion than is text. Text is the effective medium for stories, where the point is to convey emotion, but data tables and graphs are the effective medium for conveying numerical data and seeing patterns. I am trying to establish habits of thought that expect numerical quantification of daily lab activity. It seems to me, if we want students to love science and choose to be life-long learners of it, then we want science class to be filled with noisy enthusiasm, verve, modest levels of friendly competition and mathematical assessment of the functioning of rigs they have themselves made.

Introduction

The 2017 SCDS Engineering Event was a rousing success by any measure. Students were enthusiastic throughout the entire 5 weeks of the adventure. They became increasingly adept at tool use. Their contraptions showed their understanding of physical science, engineering and design. The parents who attended the community celebration of their constructions and inventions were very complementary. A good time was had by all.

Tool Use

Humans have often been differentiated from all other animals for their tool use. It seems to me that today our students spend more time at a keyboard than they do using a hammer and saw. Computer skills are wonderful but they do not replace learning to feel the heft of a hammer and the torque of a screwdriver.

Lets think about tools for a minute. Tools are what set humans on the road to civilization. They are what allowed us to become the top of the food chain. Human anatomy is rather frail. Our teeth aren't sharp like a tiger. We are not tall like a giraffe. We are not strong like an elephant. But, we are clever. And we have figured out how to use props (i.e. tools), to improve our physical prowess. Tools chipped the stone used in the pyramids. Tools wove the cloth the Egyptians wore. Tongs grabbed ingots from foundry furnaces and hammers pounded them into plows. Tools made telescopes, microscopes and Harrison's clock. Tools crafted the industrial revolution which has resulted in our comfortable lifestyle today. Although children become enamored with fantasy video games and electronic computer designs, real world tools still matter. Real world tools are still useful. Real world tools are still interesting. We no longer live on farms. Children don't touch tools today. They don't even see them. They have lost touch with their cleverness and their utility. To take a basic idea and use it in a novel way a person needs to be very familiar with it. Hand tools are our foundation. We need familiarity with them to stimulate human ingenuity and innovation. Besides, tool use is fun. There is immense pleasure in making a pile of sawdust as an elegant shape emerges from a piece of wood. The artist, the artisan, the engineer. Their intelligent hands need to feel ideas coming into existence. Children, all children, benefit from the educational stimulation of actually using tools.

Our engineering events provide the occasion for needing to use tools. But how do we organize that? What issues arise? Well, first of all, each child needs immediate access to their own set of basic hand tools. Enthusiasm falls when a person has to stand in line to wait to use a hammer. On the other hand, standing in line for the drill press works well, allowing students to observe someone using the tool and notice the safety precautions. The wheel centering station and the chassis bending jig are also communal and require some standing in line.

It is not just the need of tools. It is the collection of stuff. The organization and storage of small parts for the rig they are making. So the first thing to do is give each child a clear plastic shoebox and have them write their name prominently on it. Each child is assigned a shelf (built of cardboard under the lab stations) on which to store the shoebox. The shelf is large enough to also eventually store the assembled rig. As various parts are introduced in class discussion, students can put theirs in their shoebox until they are ready to attach them to the rig.

Oven liner foil trays are set on counters around the perimeter of the room near electrical wall outlets. Sixteen of them. One for each student. The trays are labeled, "station #1," "station #2," etc. Students select their workspace using our number lottery activity. Each foil tray has a glue gun with a tray for extra glue sticks, a 12 oz. stubby hammer,

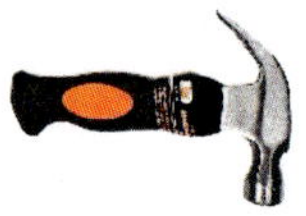

a wide reach pliers,  a lock-pliers, 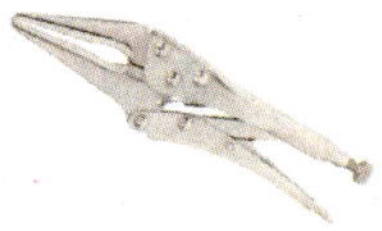

a large scissors and two rulers (of different widths). There is also a small lab stand whose base can act as an anvil. Half way through the unit, spring scales are attached to the lab stands.

At the start of the unit, students are asked to go to their station and get one tool and bring it to their writing desk. One day the hammer, the next the lock pliers, etc. Lecture - discussion introduces the tool and students practice using it on several activities. No instruction is given on the use of the glue gun but students quickly figure out how to use it (and how to deal with small burns). A tuna can is glued to the oven liner to store the glue gun so it won't be accidentally set on other things and glue or melt them.

Glue bunnies? Yes glue bunnies. But isn't that wasteful? Shouldn't there be rules? Actually, I notice that I personally have no interest in making little piles of glue to see what it will do. But a child has not explored the properties of hot glue as much as I have. So long as a child needs to make glue bunnies, they shouldn't be criticized for wasting glue. We want them to explore. We want them to push boundaries. Even the boundaries of "correct" behavior in the classroom. Just a little. Not too much mess. At some point we may need to stop them. Perhaps we can redirect them to make glue gridwork to outfit their rigs. We have to find that balance between intellectual freedom and strictly applied classroom conformity.

Our Engineering Event

Our yearly engineering topic comes from some item in the current news. It is then explored using child-appropriate mechanics and lab activity. Five weeks is a long time to engage the interest of a 10 – 14 year old child, so we usually craft three or four areas of questions related to the topic. As the weeks progress, some students remain focused on the first explorations while others who have satisfied themselves with the behavior of their contraptions, move on to the enriched level of challenges. The flurry of invention ends the day before Thanksgiving with a public display of their contraptions in the gym.

Puffer: This Year's Topic

NASA and Jet Propulsion lab have a long history of building rovers, space telescopes and even parts of the space station. Structures have to be packed compactly in the launch rockets and then deployed when they reach their destination in space. In 1987, Japanese astrophysicist Koryo Miura developed an origami fold for a solar array. In 1995 a solar panel with this design was unfolded on the Space Flyer Unit, a Japanese satellite. Since then there has been a growing interest in applying origami to all sorts of engineering. It is an amazing connection that is producing delightful ideas and clever contraptions.

JPL has experimented extensively with collapsible origami solar panels and for the past year they have been working on origami-inspired miniature explorer robots to be carried into space on their larger rovers. In March of this year they announced a robot that resembles a child's motorized toy. Even though it is quite small, about the size of a human hand, it drew crowds of delighted technology fans at a recent convention in Las Vegas.

They called it their "PUFFER" robot, an acronym for Pop-Up Flat Folding Explorer Robot. It is really a collapsible micro-rover designed to ride on board future space vehicles. NASA anticipates this miniaturized origami approach will revolutionize future space activities. They have entitled their new research program, the "Game Changing Development (GCD) program." The robot's design is in keeping with the shift towards miniaturization in our modern tech industry.

We know there are severe limits on the volume and mass that you can send into space on any one rocket. NASA has had to get creative with landers and rovers, making them increasingly small in size, light in weight and foldable. Although Puffer is already small enough to be an excellent option for exploring some of the harshest landscapes in the solar system, there is talk of adjusting its size up or down. Some engineers want larger instruments, more batteries and solar panel area, or possibly larger wheels for mobility and ground clearance in rougher terrain. Some at JPL want to add a number of instruments that would allow it to sample water for organic material, or study the chemical makeup of its environment. Others are seeking to scale Puffer down to pack more units into the existing launch rockets or pack more on the next generation of landers.

Puffer is a speedy little guy. On a level dirt path, it can drive about 2,050 feet on one battery charge. If Puffer needs to recharge, it can simply flip over and collect sunlight through solar panels on its underside. A tail was added to assist in the flipping so it can recharge in the sun. With this recharging system, engineers expect Puffer to operate for months and possible years at a time. The mobility was designed by a student who had worked at UC Berkeley's Biomimetic Millisystem Lab, where he worked on developing robotics based on natural forms, like animal and insect movement. Robot bugs at Berkley developed a "skittering walk" that keeps the bot inching forward, one wheel at a time, without slipping. The "skittering walk" provides flexibility to adapt to what it encounters. That process was incorporated in the way Puffer moves.

Originally, it seems, Puffer was designed by combining various origami techniques and biomimetic movements. Its two wheels bounded a circuit board chassis which collapsed inward following origami folds, ending up flat rather like a horseshoe crab. As the project developed, origami was replaced with 3D printed parts and circuit boards. But the flat-folding attributes remain.

The most striking feature of the Puffer is that it is a collapsible micro-rover. It has two foldable wheels with prominent treads, a tail which provides stabilization, solar panels which can easily be flipped upward to face the sun, and complex electronics, which allow it to be wirelessly controlled. The flattening maneuver has two functions. Several Puffers can be flattened like cards and stacked one on top of the other to be carried inside a standard rover. Once deployed Puffers can flatten or straighten their wheels to meet environmental conditions. It can tuck in its wheels and crawl into small crevasses where normal rovers can't fit.

The hinge material for the origami folds was particularly troublesome. The folds had to able to repeatedly collapse without damaging the copper wires connecting the electronics. The Puffer needed to be able to survive 5,000 cycle tests. The underlying material had to remain unfrayed. They ended up using a material called Nomex which was used in in the air bags that cushioned NASA's Spirit and Opportunity rovers when they touched down on Mars. Nomex is also used in firefighting to repel heat. Engineers believe Puffer could survive the punishing high temperatures sometimes encountered in space.

The excitement over the fold-down aspect of the robot comes from anticipation that it will be able to crawl under ledges and into caves that were not possible to explore before. These tiny robots could crawl across sand dunes where normal rovers would sink in. They might explore lava tubes, deep caves or overhanging rocks. Although they might venture too far to maintain communication with the mother ship, a line of them could relay data back as needed. They might explore icy slopes of craters where the footing is bad. Much of the surface of Mars is boulder-strewn and treacherous with sudden crags and cliffs. Large, expensive rovers can't take the risk of exploring such "chaos terrain." Puffers are designed to fall off cliffs (as much as 3 meters in the Martian gravity) and right themselves like a cat. The folding chassis provides a lot of flexibility and impact absorbing capability. Uses on earth might include dropping Puffers from helicopters to explore remote inaccessible terrains.

The special spikes on the wheels allow climbing and crawling over unusual surfaces. When the wheels were printed on 3D printing machines, these micro-spines were added which remarkably improved climbing ability. Puffers now can climb 45° slopes whereas most vehicles are limited to 30° or so. This unique mobility would allow Puffers to drop into crevasses and climb their way out. When in trouble, a group of tiny Puffers could help each other and work in tandem the way rock climbers do.

Once a Puffer managed to get into an inaccessible space, it would use instruments to assess the area for its water history and habitability. Instruments such as micro-imagers and tiny spectrometers are being developed. An Illinois optics company is developing a microscope the size of a sugar cube to attach to Puffer's underside near the solar cells. It is hoped that it would capture images of the soil as it skittered along. Additional work is being done to develop high resolution imagers that could see (and transmit images of) objects much smaller than a human hair.

The culmination of the development of a tiny hand-size robot is to make a bunch of them that can communicate to each other. A rover could eject a small swarm of them like baby seahorses escaping from their parent's belly. They could communicate with each other and do parallel science. In addition to helping each other deal with precipices and escarpment, a swarm of Puffers could enormously increase the amount of data collected in a single day.

Many potential applications of tiny Puffers are under consideration. Of course they would work well on Mars and moons of other planets. They might scout the possibility of mining on asteroids. On earth, they might track the fluctuation of ice in polar regions or be used for disaster relief search and rescue. The possible future of origami micro-robots seems endless.

Origami Engineering

Origami engineering seems to be the wave of the future. If one has a problem that involves collapsible folding structures, origami may be the answer.

John W. Hetrick obtained the first automobile airbag patent in 1953. which became one of the first applications of origami folding to take hold worldwide. Even before that, as far back as 1943, folded honeycomb structures stiffened with resin were used in aircraft construction. Today such honeycombs are standard fillings for panel doors. The vast majority of practical applications have come within the past 50 years. Advances in number theory, computer science, and computational geometry have paved the way to extend origami principles far beyond the art itself. Large and small, from buildings to heart stents, origami is reinventing the mechanical world. Folding kayaks, tents and refugee shelters are widely publicized examples. Even backpacks and clothing are feeling the influence of origami – the Japanese art of paper folding.

Much of the experimental engineering work being done at universities employs the Miura fold. This basic fold, invented in 1970 by astrophysicist Koryo Miura for use in solar panels, has inspired so much mechanical invention that in 1981 the British Origami Society officially named it the Miura-ori. It consists of successive creases that are either mountains or valleys. At the next vertex there is a reverse of the fold. Mountain become valleys and valleys become mountains. These folds are used to create accordion and corrugated patterns used in many new and useful applications. It is often made into a flat platform but once you get the hang of it, it can be made into a curve or other shape. Artists have sculptured amazing vases and statues out of tightly folded Miura folds.

This fold is currently the most commonly used crease pattern in engineering applications. It is particularly helpful for impact force distribution and dissipates energy through crushing or plastically deforming its shape. Packages using this fold can survive being dropped from helicopters in disaster areas or in delivering rovers to other planets. The platforms are strong. They can support their own weight along with other loads. That allows this fold to be used in designing solar sails, roof structures, antennas and survival tents.

Some origami engineered designs are static and some are active. Some are deployed one time and others fold and unfold repeatedly. Delivery and storage structures usually deploy just once. Heart stents and artificial organs unfold once. The Puffer wheels are designed to fold and unfold five-thousand times.

A Force To Move It

We made split axles so the wheels fold up. We made Miura-ori platforms on the chassis. But the rover just sits there. How can we cause it to move? We can use rubber band power for the drive wheels, but what can we use to stretch and shrink the origami box holding the wheels? How about a syringe system? Pneumatics! We have to learn about them both. Rubber band power and pneumatic syringes.

Rubber Band Power

We, as adults, know that we have to stretch a rubber band to get it to move a toy car. We have to accumulate potential energy. A 4th grade student may not know this. I have watched them neatly wind a rubber band around an axle (like a spool of thread) over and over. They set it on the floor and nothing moves. They try again and again. Finally, they watch another child who is a "local expert" wind their car to move it ten floor tiles. After many false starts their fingers start working. They learn by doing, watching, being frustrated and trying again.

Pneumatics

And what about pneumatics? Do we present it in the standard way discussing compressibility and coefficients of expansion? As you might guess, we do not. How can we cause children to raise important questions about pneumatics? What kind of introductory activities help them search for answers? What kind of numerical data collection, (from each lab in the room) can help them see trends and generate more questions. And be playful fun? I hung a spring-scale on the lab stand at each glue gun station and asked students to record the force registered to pull the plunger on a 3ml syringe. Then I gave each child two matching syringes and a foot-long aquarium air tube. As I usually do, I carefully watched the children as they figured out how to connect the "push-me-pull-you" system. They had to figure out how to stretch the tube using the hot glue gun so it was large enough to fit the syringe neck. Then they had to discover how to position the plungers to get the system moving. In the process someone noticed the "glide" of one stopper after the other end had stopped pushing. Great! A chance to collect data. How far does it glide? Does the position in the tube matter? Pop! Pop! Very soon enthusiastic hands popped the plunger out. Rather than admonishing them to be careful, I said, "OK, lets measure how far they can launch." (I was careful to say launch rather than shoot.) And I said if they wanted to test that they had to "clear a course" so no one got hit. Oh yes! Everyone was excited to play with that idea for quite a few minutes

– and they dutifully recorded their own data. The next day we shared data from each person in each of our journals and discussed uncontrolled variables.

As soon as they were satisfied with their discussion, I said, "How about a 6ml system? Students enthusiastically scurried to lab to assemble their 6ml "push-me-pull-you" and collect the three sets of data: spring scale pull, glide inside syringe case, and launch distance. As usual they knew that each test had to be repeated three times to have confidence in their data.

Again the following day, after taking class data and discussing it, a pair of 10ml syringes and a foot-long tube were given to each child. The following day it was 20ml syringes and finally 60ml syringes. By this time every student had in their journal a nice recording of class data for each type of syringe. Now it was time for mixing and matching and making theories about how a projectile could launch even farther. Try it! Record it! Laugh and try another! Play with the process of science. Come to love the scientific way of knowing and understanding what you know!

Remembering Children's Contraptions

Children tinker. They build real contraptions. Just like an engineer.
That's what makes it fun.
Designing.
Building.
Repairing and revising.
Taking data.

Trying over and over.

Now Origami.
That seems different.
We made a paper crane in the second grade, but it just sat there.

There was nothing to test.

Origami is pretty. It is art. Not much like engineering.

In March, we read about a new robot NASA had made.
It was called the "Puffer."

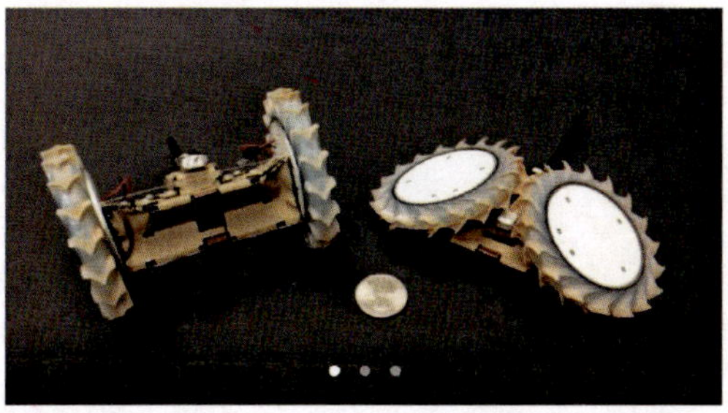

Pop-Up Flat Folding Explorer Robot (PUFFER) was inspired by an origami design. This little robot could be used as a scout for larger rovers, going places that would be risky or hard to reach. Image Credit: NASA/JPL-Caltech.

Pop-Up Flat-Folding Explorer Robot.

Hummmmm.
That seemed interesting.
It folded. It moved.
It had something to be tested.

Puffer could flatten to crawl under ledges.
Puffer could pop up to climb over rocks.
Puffer was a moving origami mechanism.

We thought about that.
How could you make a pop-up origami mechanism – out of junk?
Out of the junk we have?
We looked around at our stash of odd materials.

We needed something stiff that we could fold.
Shiny would be nice. NASA stuff was usually shiny. Like metal.

We found a pile of silver colored cardboard lids. Deli lids to take home apple pies.
There were lots of them.
Enough to try and fail and try again.
If you scored them, they folded nicely and held their shape pretty well.

OK. Pie lids. The silver color was pretty.
We thought about origami.
To make a blow-up origami ball, you had to make a box.

How could you make a box out of a pie lid?

Well, first you had to find the center.
We know that the perpendicular bisector of chords of a circle go through the center.

We drew two chords and made their perpendicular bisectors.

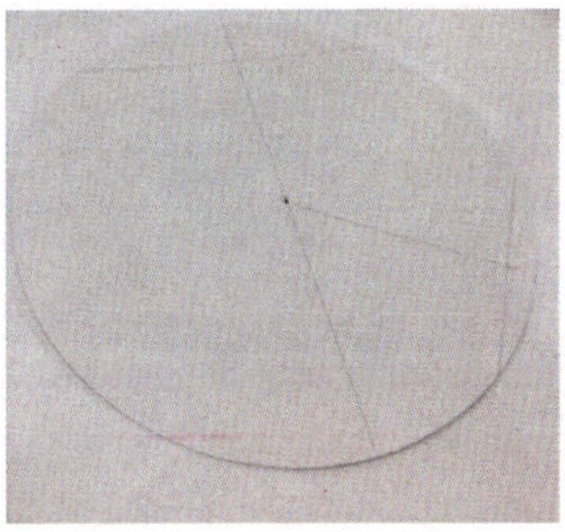

Then we drew a third chord and bisected it just to check our findings.
The intersection seemed to be the center of the circle.
We could spin the silver disk around that point and it didn't wobble.

We got two rulers from our glue gun stations.

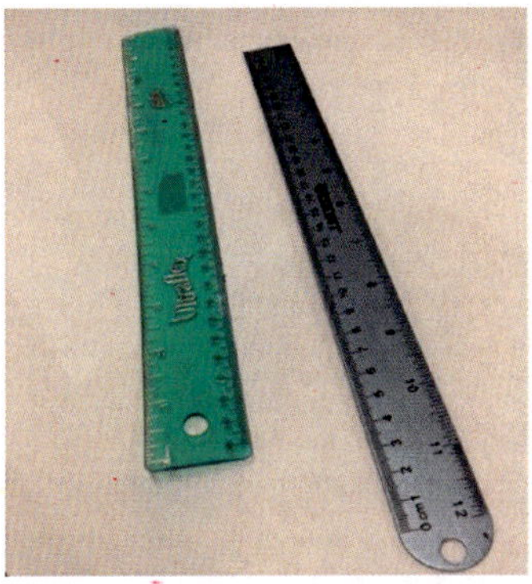

One was wider than the other.

We drew a line across the circle through the center.
Setting our wide ruler carefully we drew lines a ruler width away on either side.

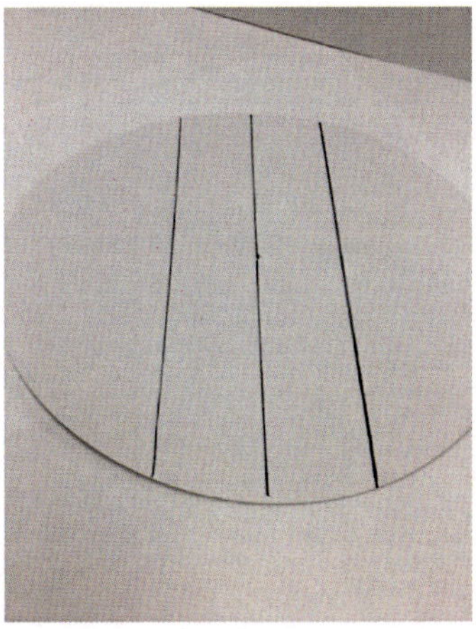

Then using our thinner ruler, we drew a set of lines outside of that.

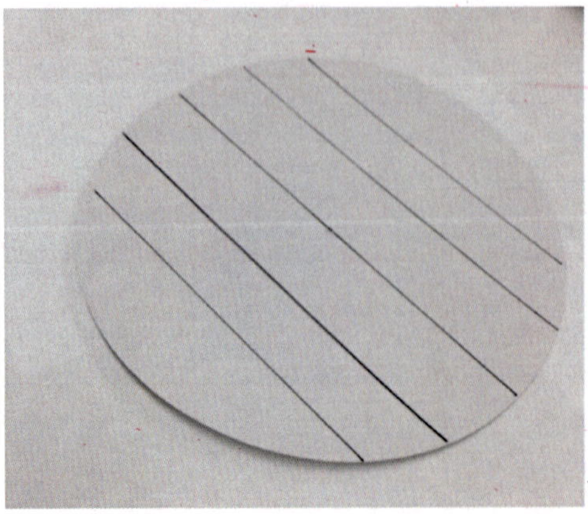

We were careful to keep everything parallel.

We cut off the half-moon shapes outside the lines on either side.

That made four "pointy" places like corners.
We connected the corners with a line across each end.
After we cut off the half-moons at the end we had a rectangle.

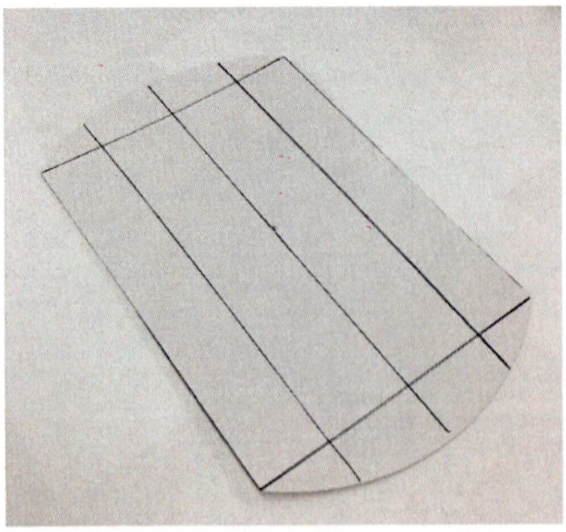

Next we drew a line (using the smaller ruler) parallel to each end.

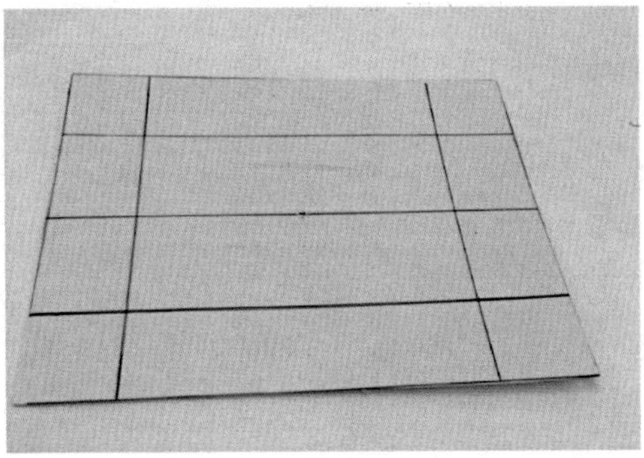

At each end there was a little square.
We drew a diagonal in each square pointing out toward the corner.

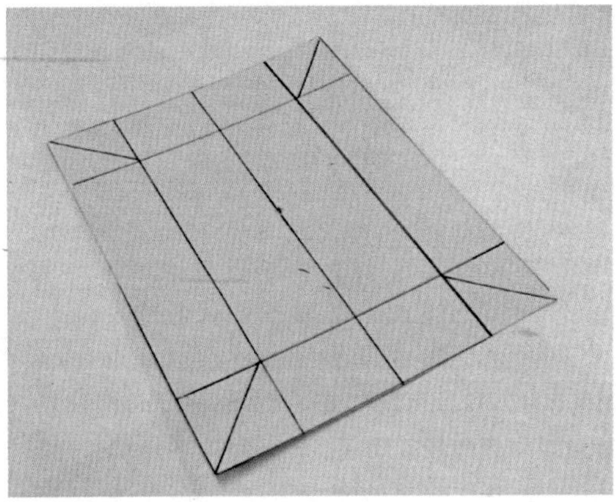

The silver cardboard was soft enough for a pencil to score it.
We scored every line ten times. But we were careful not to score the line that went through the center.

We then began to fold the scored lines.
Back and forth until the paper wanted to move on the straight lines.

At last we folded the box up – with "bunny ears" sticking out at each corner.

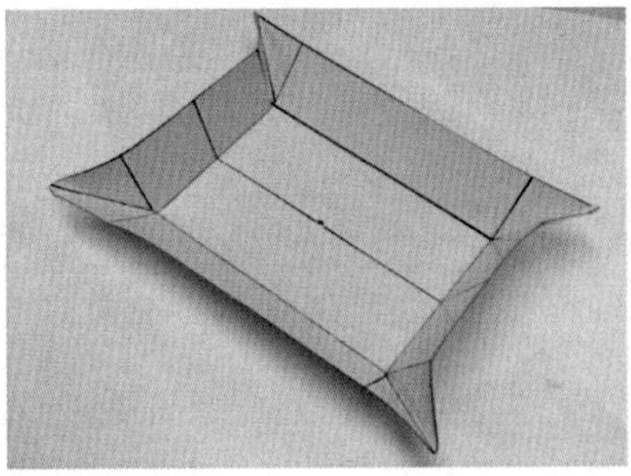

We folded the “bunny ears” around the ends and glued every corner. The glue gun did the job perfectly. We didn’t get burned very much.

After that we held the hot corner with a wrench while the glue dried.

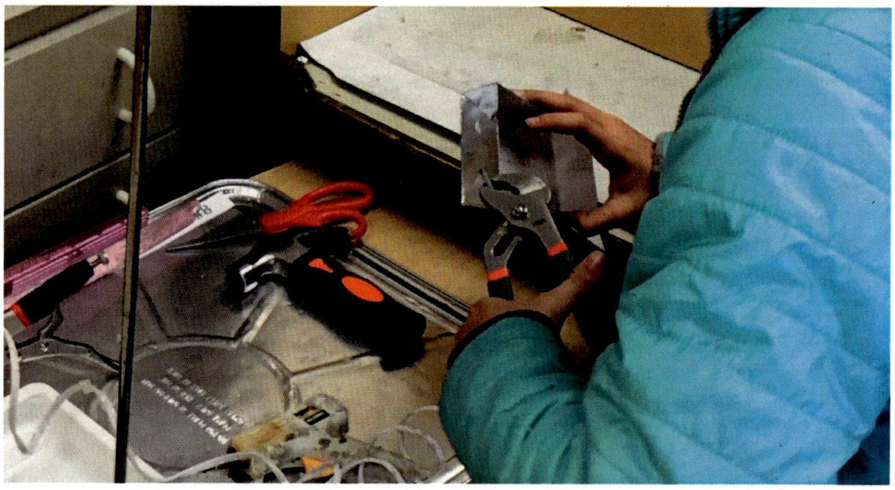

Sure is nice to have tools.

What a nice looking box!

But will it fold down like the "Puffer"?

We found we could push all the sides in and fold the corners carefully so the ends lapped over the sides. We could make the box pop-up and fold-flat.

But it wasn't like a NASA rover .... Yet!

It needed wheels!

We made two wheels with short axles like the Puffer had.
They were hard to make because everything was so small.
We held the axles with pliers and used our lab stand as an anvil as we pounded on wheels.

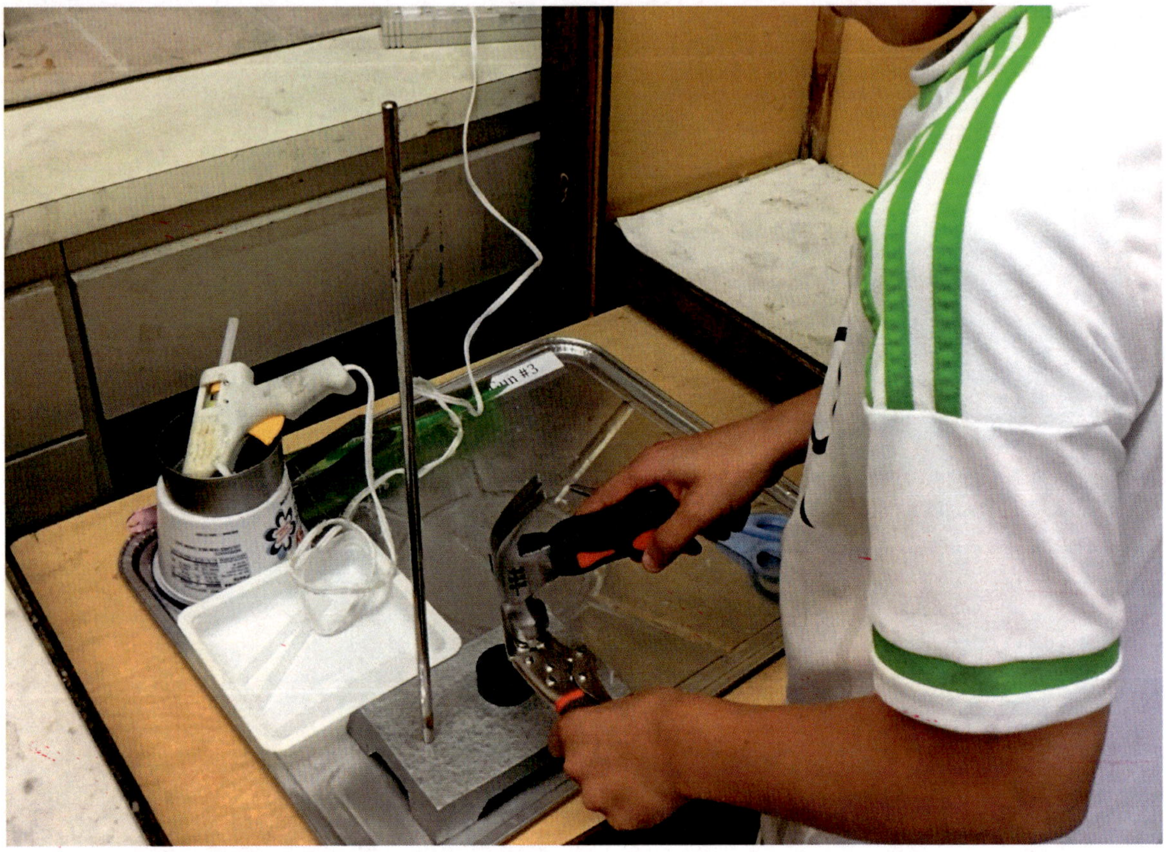

The axles went through the wheels but there were problems.

How do you hold the axles so the wheels will turn?
Oh yes! We need bearings. Straw bearings!
But with these short axles, everything will fall off.
Little metal axle retainers could be pushed on to hold everything in place.
Retainers are soooooo tiny.
How could we keep from smashing our hand with the hammer?
Tools again!
Put the wheel on the axle and use it to prop the axle upright.
Set the retainer gently on the top of the axle.
Set the lock-pliers on top of the retainer and gently hammer.
With practice the pliers can push the retainer down.

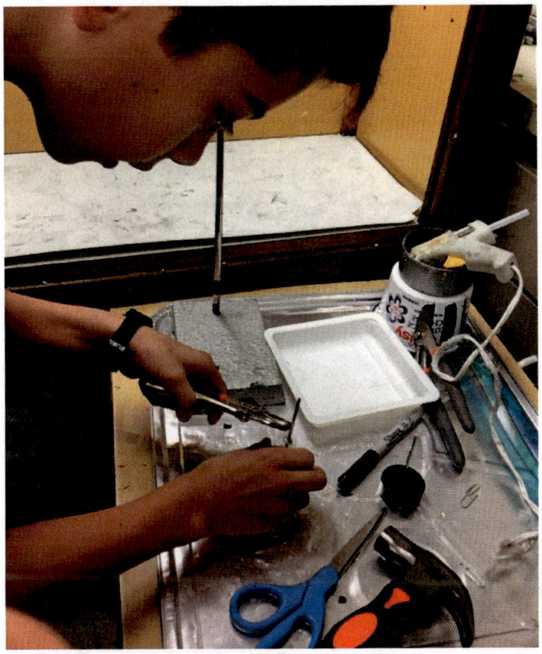

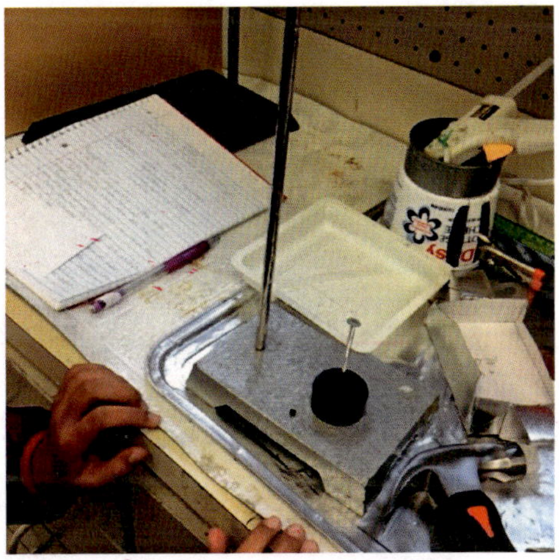

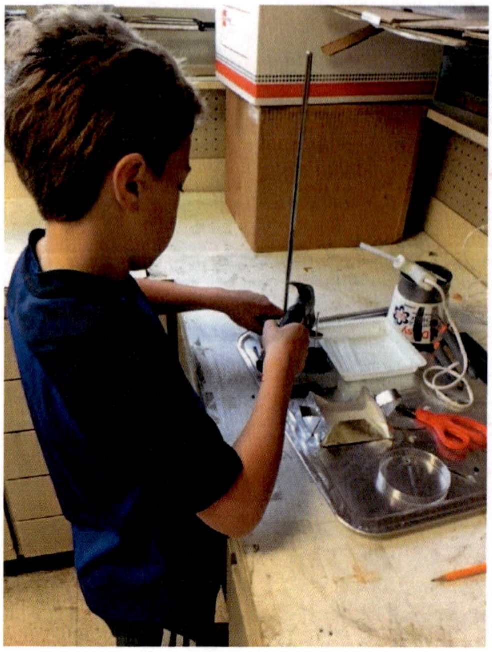

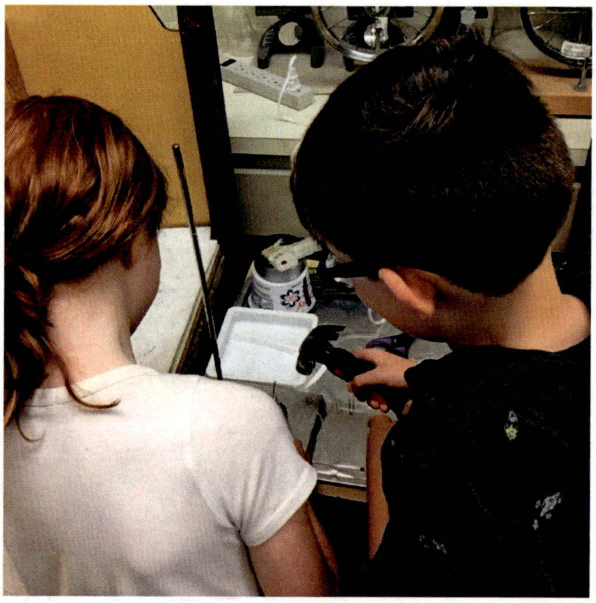

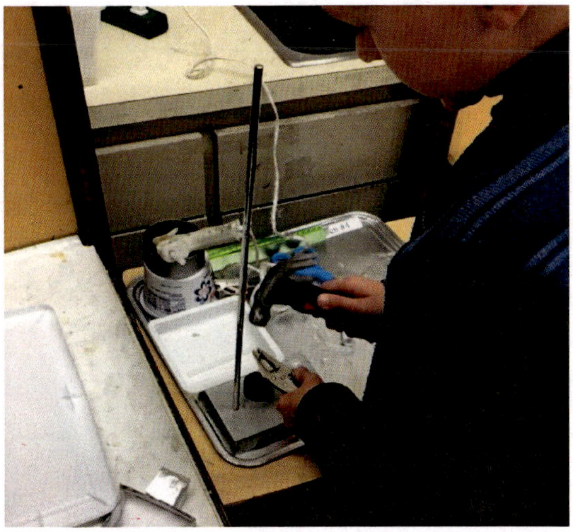

Retainers worked.
The straw bearings stayed in place.

Great!
Now how to balance the axles on the box ends.
That seems a little wobbly.
Hummm.
Students looked around the room.
We noticed Ikea shelf brackets holding shelves on the wall.
Maybe we could put shelf brackets on the box ends.
The left over silver half-moon strips should work just fine.

We wrapped the strip around a ruler and slid it off to make a triangle.
We slid the ends apart until one angle was ninety degrees.
Just like a shelf bracket.

We glued the shelf brackets to the box ends.

Now we just had to glue the straw bearings to the little shelves.
OH NO! There was a problem! The wheels won't touch the ground.
But we could solve that.
Just use bigger wheels!
Soup lids!
Just glue soup lids to the black plastic wheels.
Soup lids should be big enough to touch the ground.

It is helpful to find the center of the soup lids so the axle fits through.
Cottage cheese lids just fit inside and they have a center bump.
Just stab through and you have a center hole for the soup lid.

Would it work?
Would the big wheels on the little shelves fold down? Pop-up?
It seemed to work pretty smoothly.

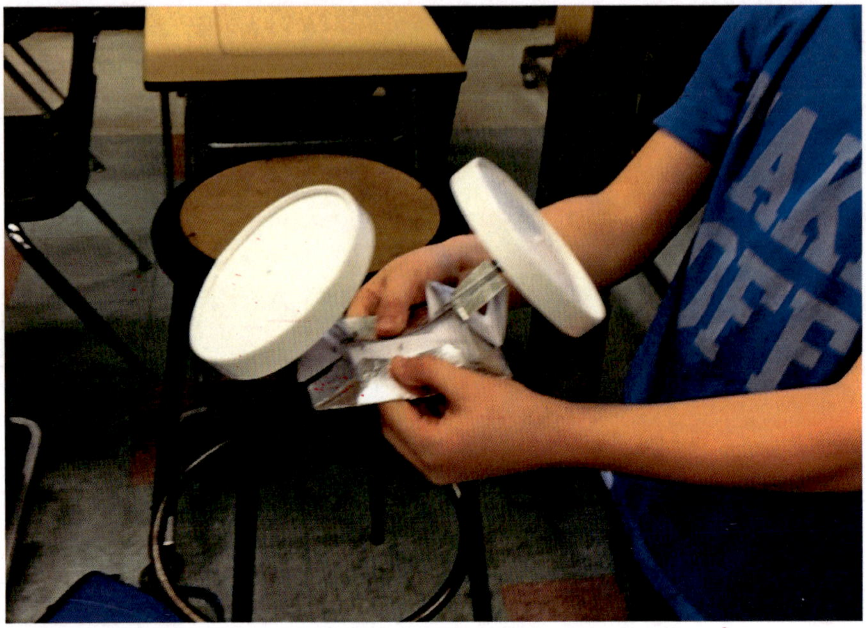

We all showed our clever pop-up rovers to our friends and soon lots of people had made their own. They popped up and down easily.

People began to sing
"Puff, the pop-up rover..."

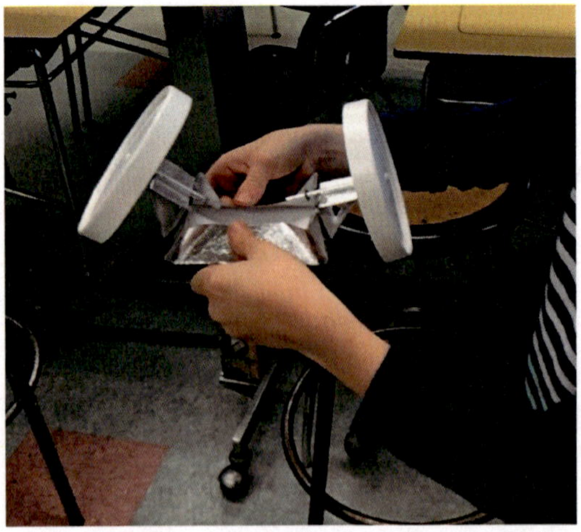

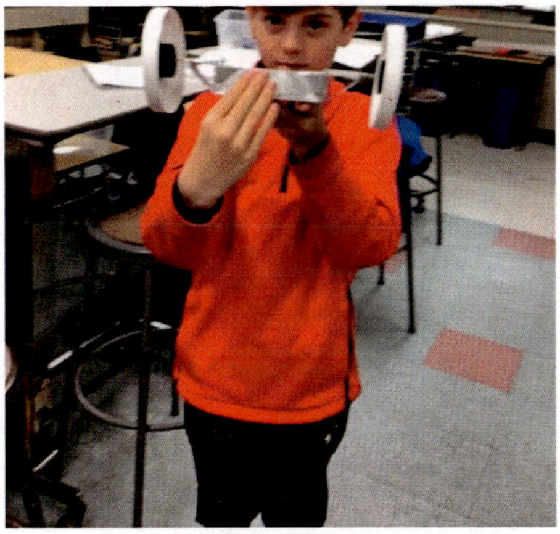
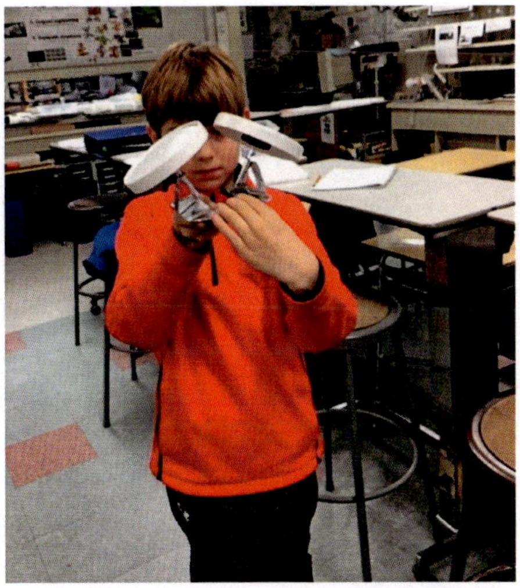
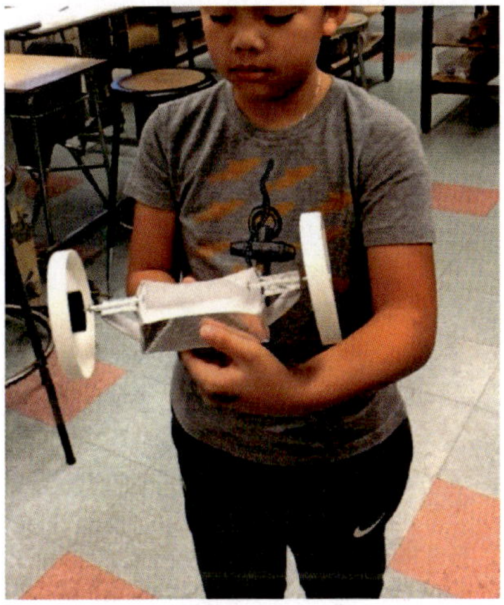
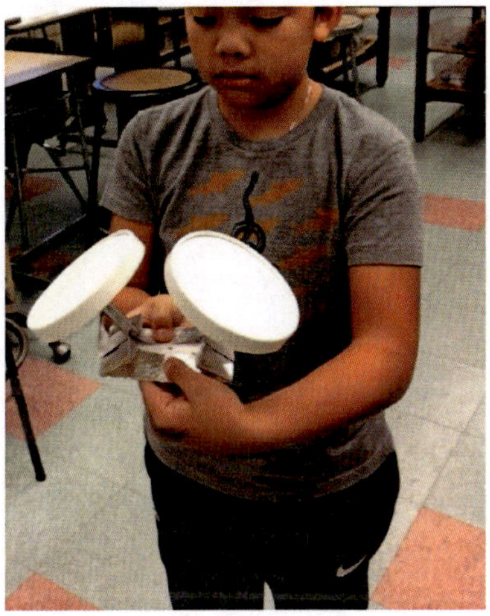

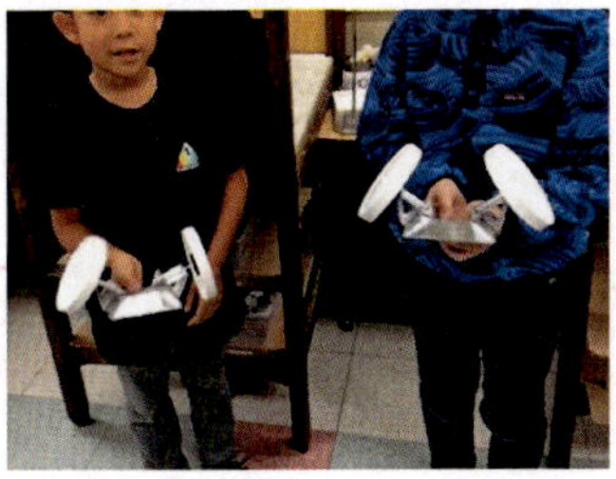

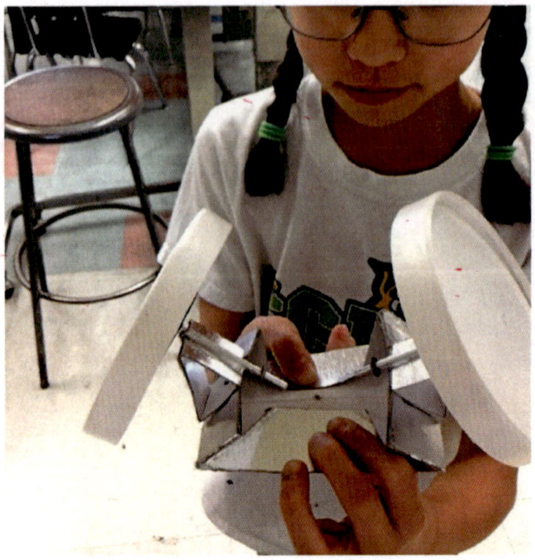

They sat nicely on the floor.

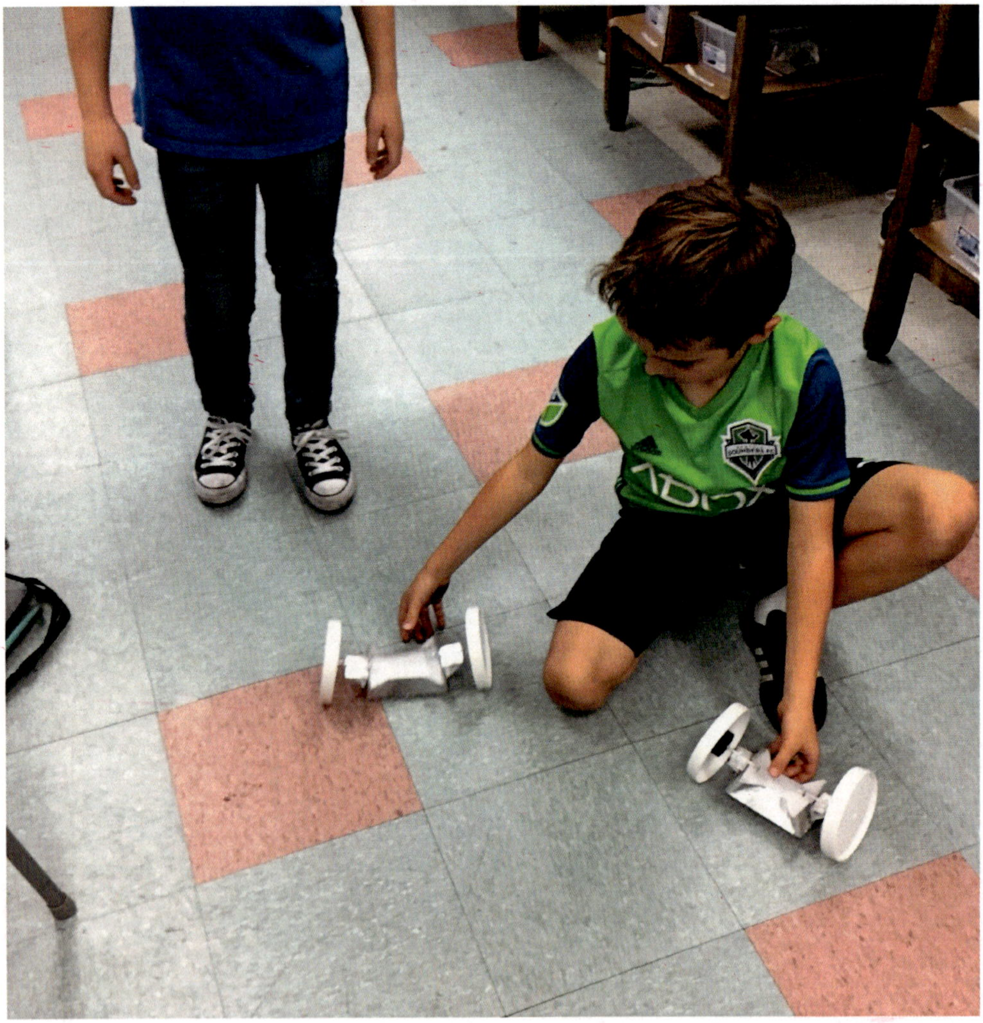

Sometimes the glue came lose and repairs were needed..

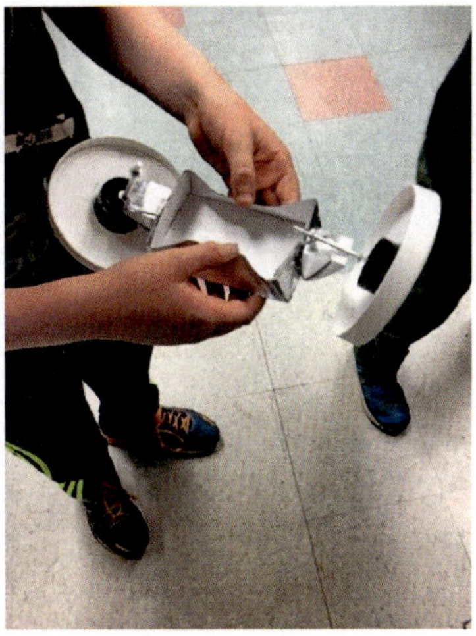

A Chassis To Drive it

But there was a problem.
The NASA Puffer had an electric motor on each wheel.
We didn't have any electric motors.
We did have lots of rubber bands left from model airplanes.
Now, if there were just some way to put them on our pop-up boxes,'
We could race them down the room.

Rubber bands need to stretch and then unwind.
The Pop-up wheels needed a chassis.

We had a couple of packs of ceiling-hanging wire.
We could cut it in half and then bend the three-foot lengths into chassis.

We screwed four screws into the end of a long board and glued tongue-depressors in the middle. The bending-jig had to be about as wide as the Pop-up box so everything would fit.

You needed a rear axle with a set of wheels.
And, of course, straw bearings so the axle would turn.

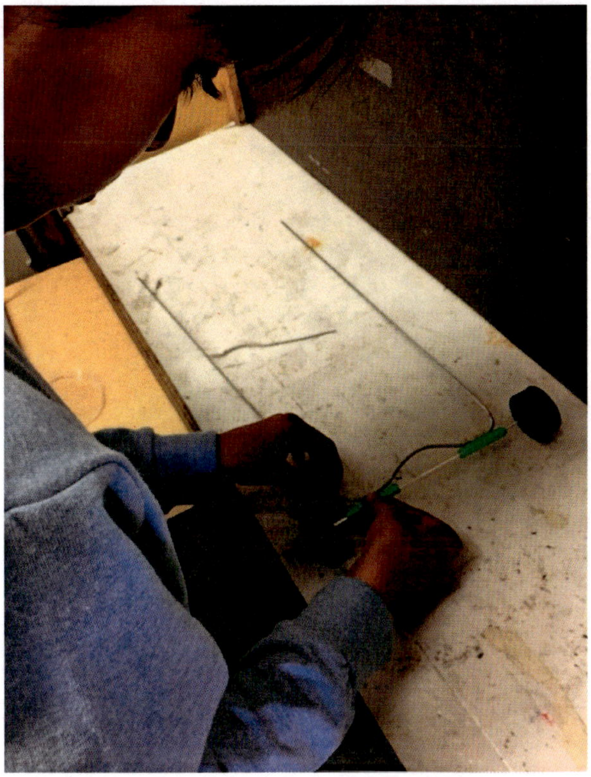

Then you had to poke the ceiling wire chassis into the folding box.

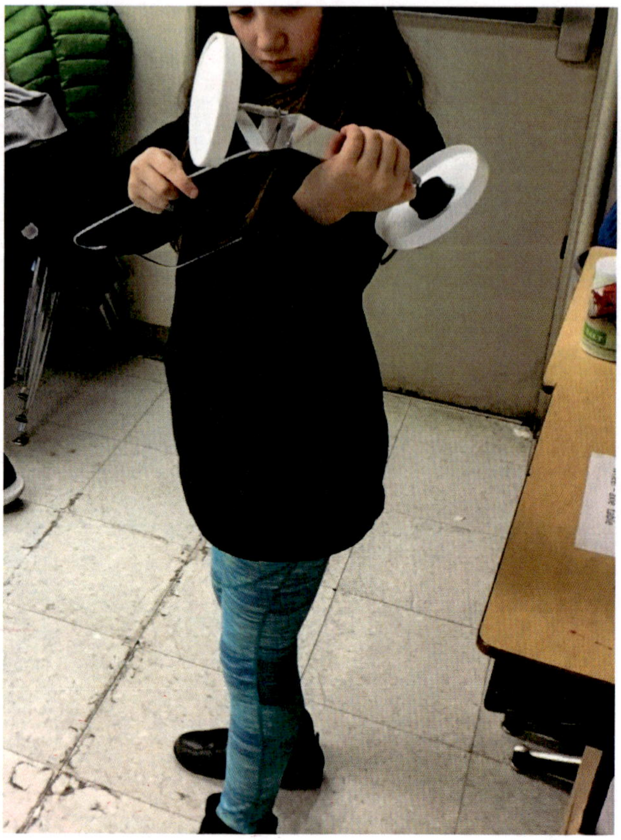

And finally you could attach a rubber band to see if the contraption would move.

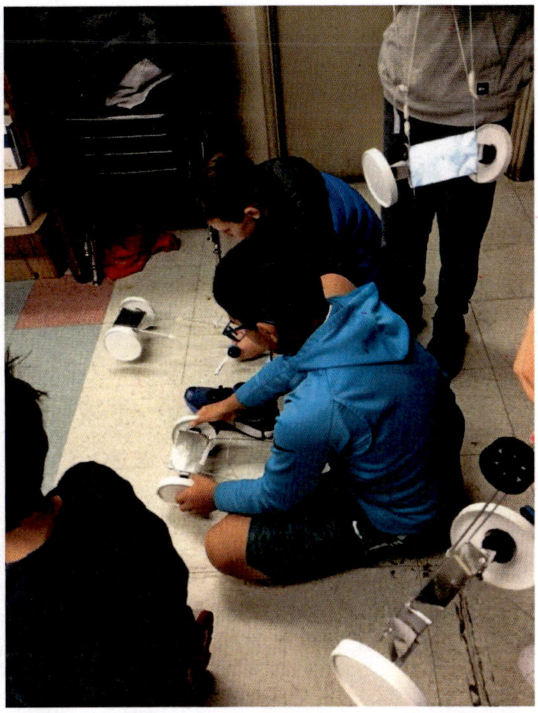
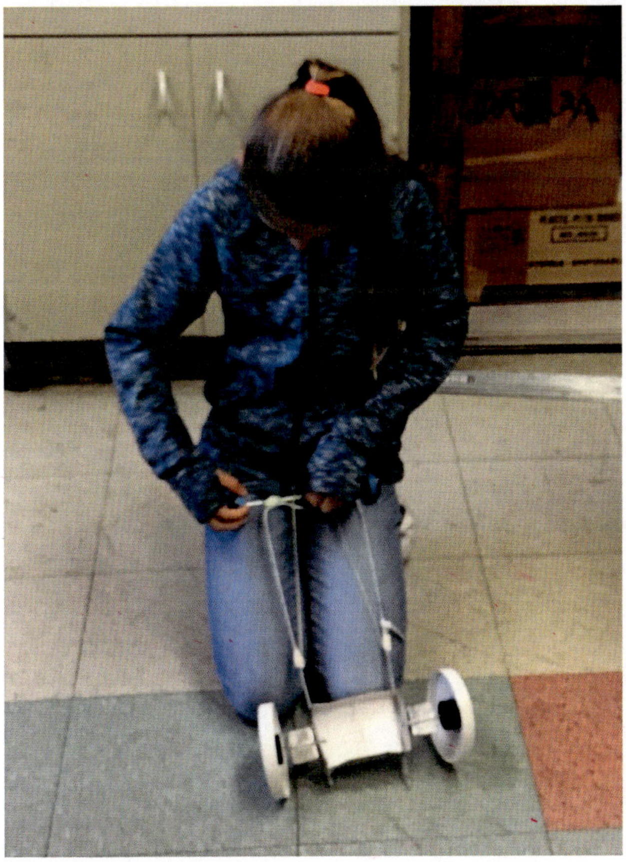

Names went up on the board!
Our rigs were moving.
The records show they went about ten floor tiles.
Could we make them go farther?
Winding the rubber band carefully helps a lot.
But maybe bigger wheels would make it go farther.
Soup lids!
A lot of people rushed to glue soup lids to the rear wheels

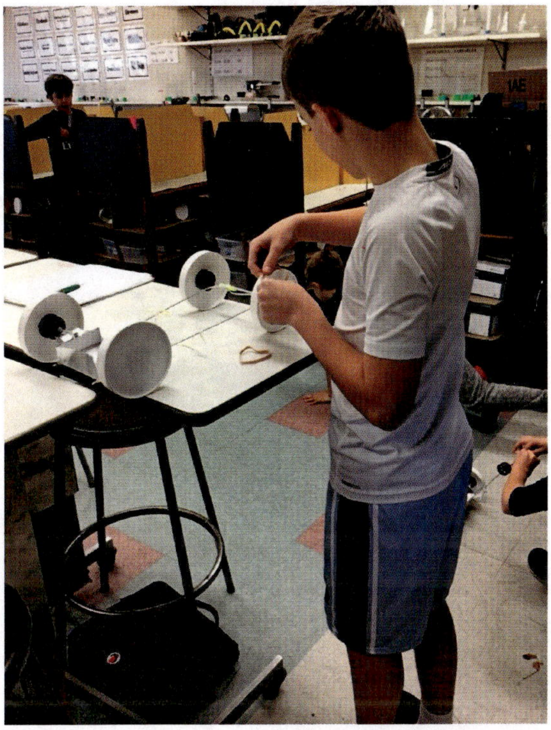

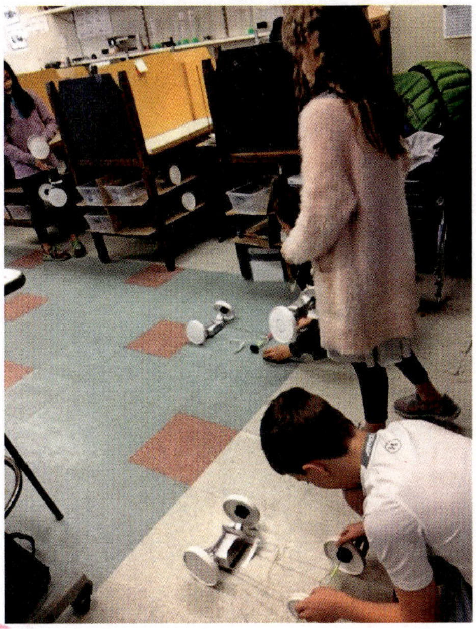

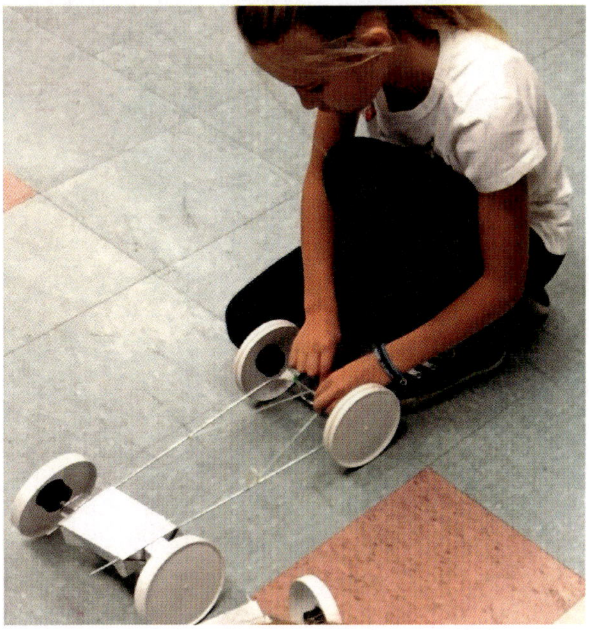

Some people popped off the big drive wheels and tried the small ones again.

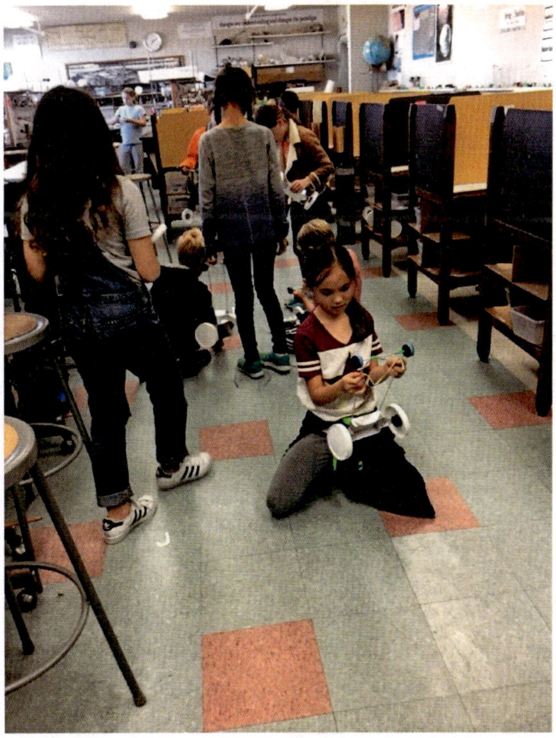

Pistons To Deploy The Mechanisms

Rigs were zooming down the length of the room.
Twenty-five, thirty-five floor tiles.
Big wheels really went a lot farther.

But there is one more thing we have to do.
We have to figure out how to mechanically raise and lower the folding wheels.
To make them pop-up.
We need an on-board mechanism.
We can't reach down and deploy them by hand.
What can we use?

Up and down. In and out.
Folded flat and fanned open.

That sounds like a piston to me.
Can we use pistons?
Hydraulics?
Pneumatics?

Well, no water in the gym. No water leaks in the science room.
That means we can't use hydraulics.
But pneumatics, air filled pistons, would be fine.

How do they work? How big are they?
How can we test them to see what they can do?

Let's start with 3ml syringes.
Every person gets a set of two syringes and a foot-long connecting tube.

Oops. That doesn't fit.
Wipe the glue off the glue gun and insert it in the end of the soft tubing. The tubing will stretch. You can quickly slide the soft hot tube on the syringe end.

A syringe on each end of one tube.
Back to back.
Like a "Push-me-pull-you."

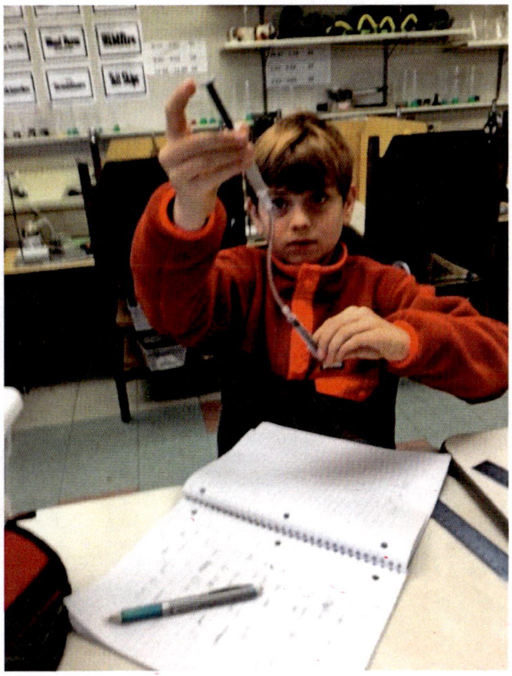

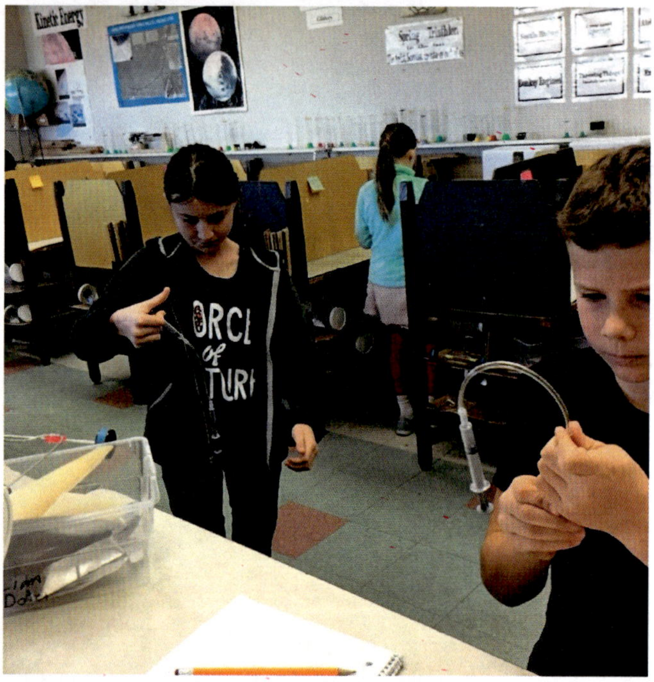
Kinetic Energy

DYNAMO
adidas
MLS

Data tables!
We now have force measurers at each station.
How hard do we have to pull to make the plunger move?

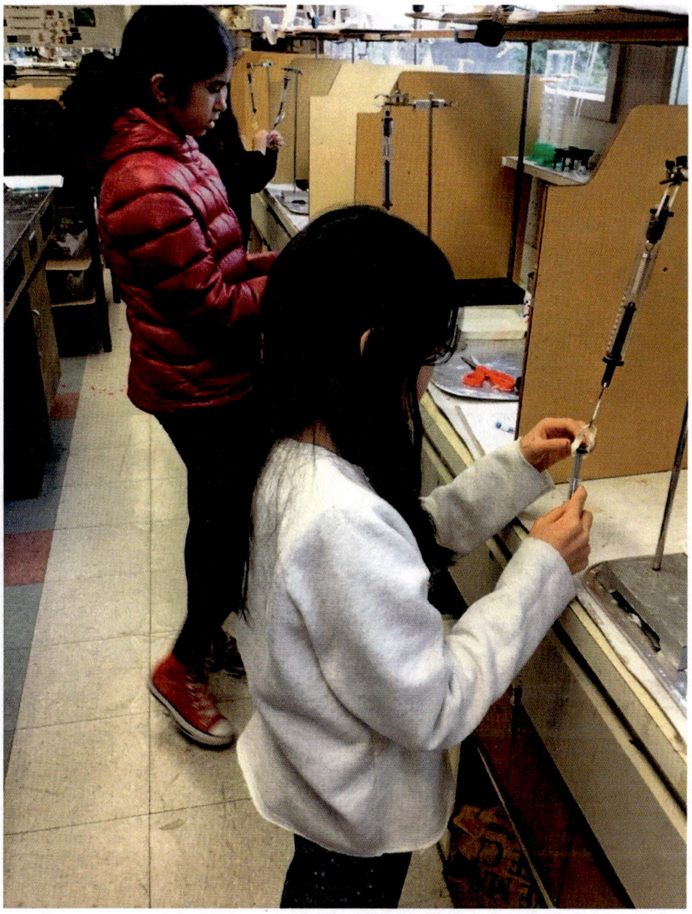

Compression. How far does the plunger glide when the push stops.
But mostly launching.
Shooting the plunger across the room.
Pop! Pop! Pop! We can get data and have a funny time doing it.

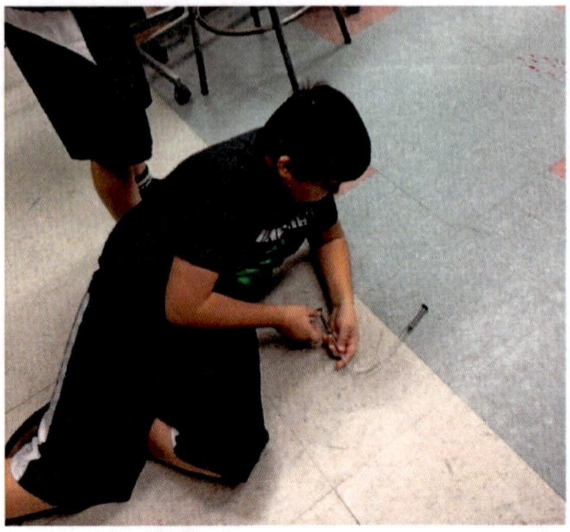

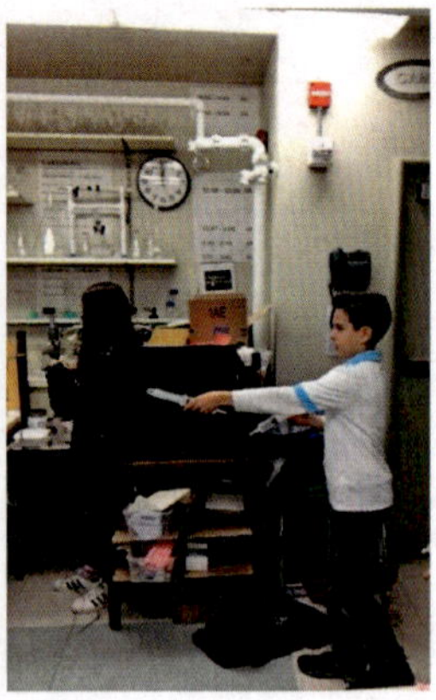

Each day we get a new size.
A pair of syringes and a tube. To keep.
We each get to keep a full set.

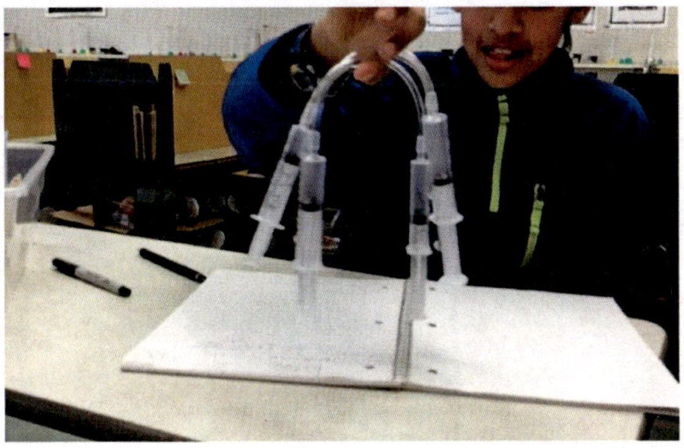

Two 3 ml and a tube.
Two 6 ml and a tube.
Two 10 ml and a tube.
Two 20 ml and a tube.
Two 60 ml and a tube.

There seem to be trends in the data we are gathering.
As tube size gets larger the force-measurer shows it takes more force to move the plunger.
Bigger tubes often seem to shoot farther.
There seem to be a lot of uncontrolled variables.
Friction matters a lot.

So, with all this data, we can predict how to mix and match them.
Get another tube and try a mismatched system.
Try to discover how to power your pop-up box.

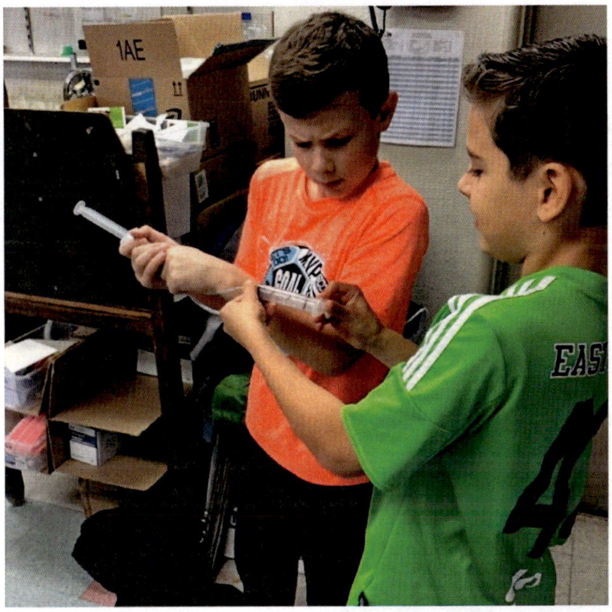

Origami Platforms

Well, this isn't very good.
When a syringe is attached to the box mechanism, the other end drags on the floor.
We need a platform!
A platform to set the syringe on when the rig zooms across the room.

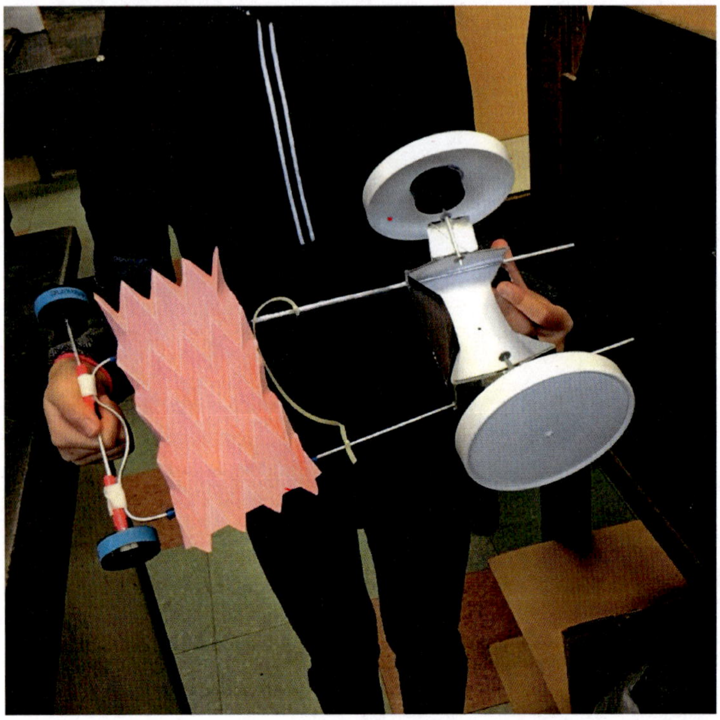

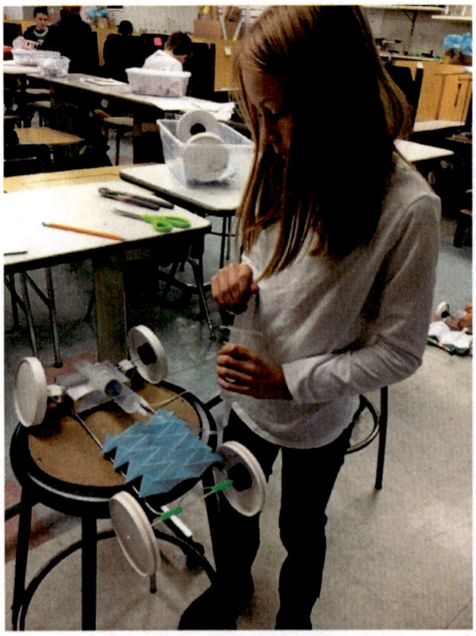

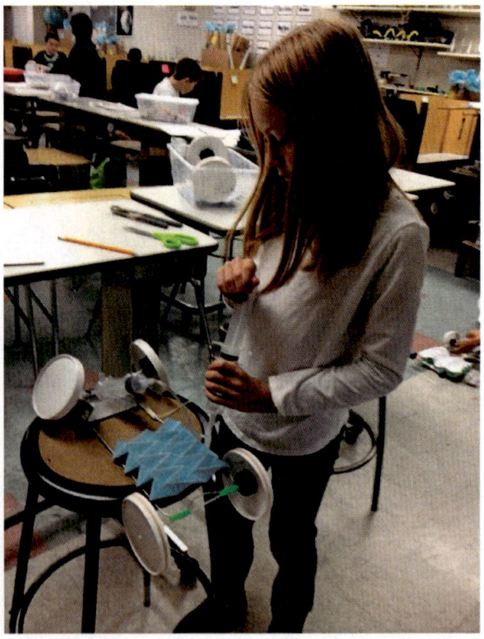

But, such a nice platform can carry other stuff.
We can make origami huts that collapse in the same direction as the platform.
That way they can move together.
It is good to make small packages for launching things into space, and then be able to expand them when they arrive at their destination.

We have a lot of syringes. We can use combinations of them to operate lots of stuff.

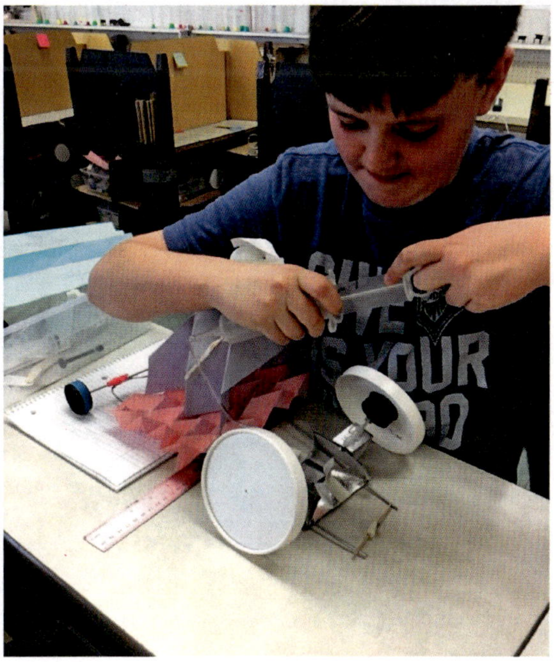

Games And Deployment

One good idea leads to another.
A hut on an origami platform can carry a ping pong ball.

Fold the origami into a claw.
A 3x5 card can make a claw to carry a ping pong ball.

Tall Ships

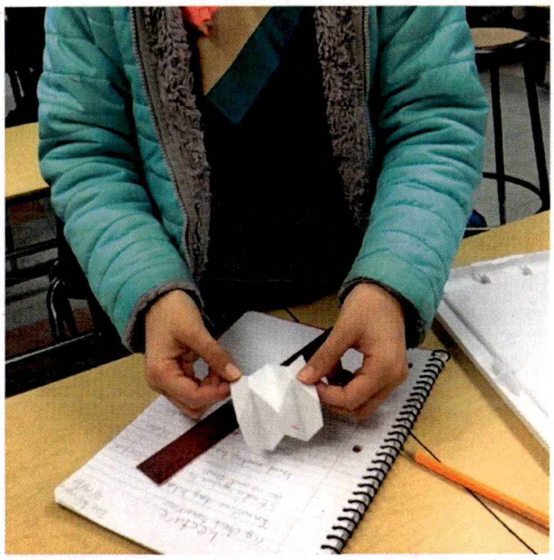

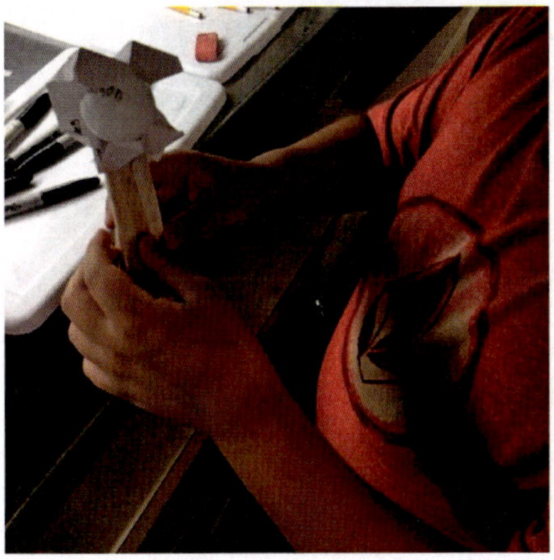

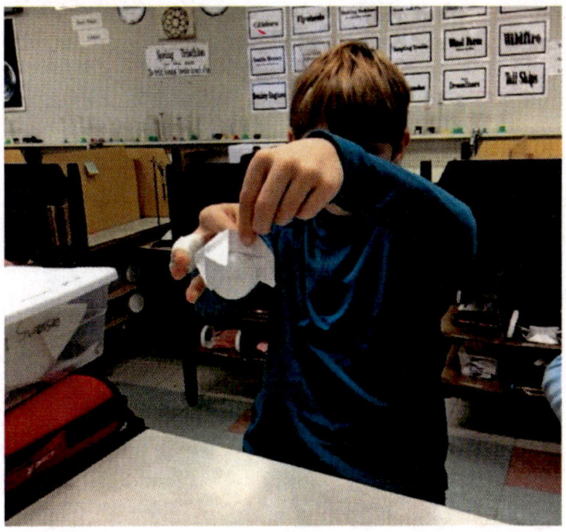
Wildfire
Tall Ships

We can make the claw open and close using our hydraulics.

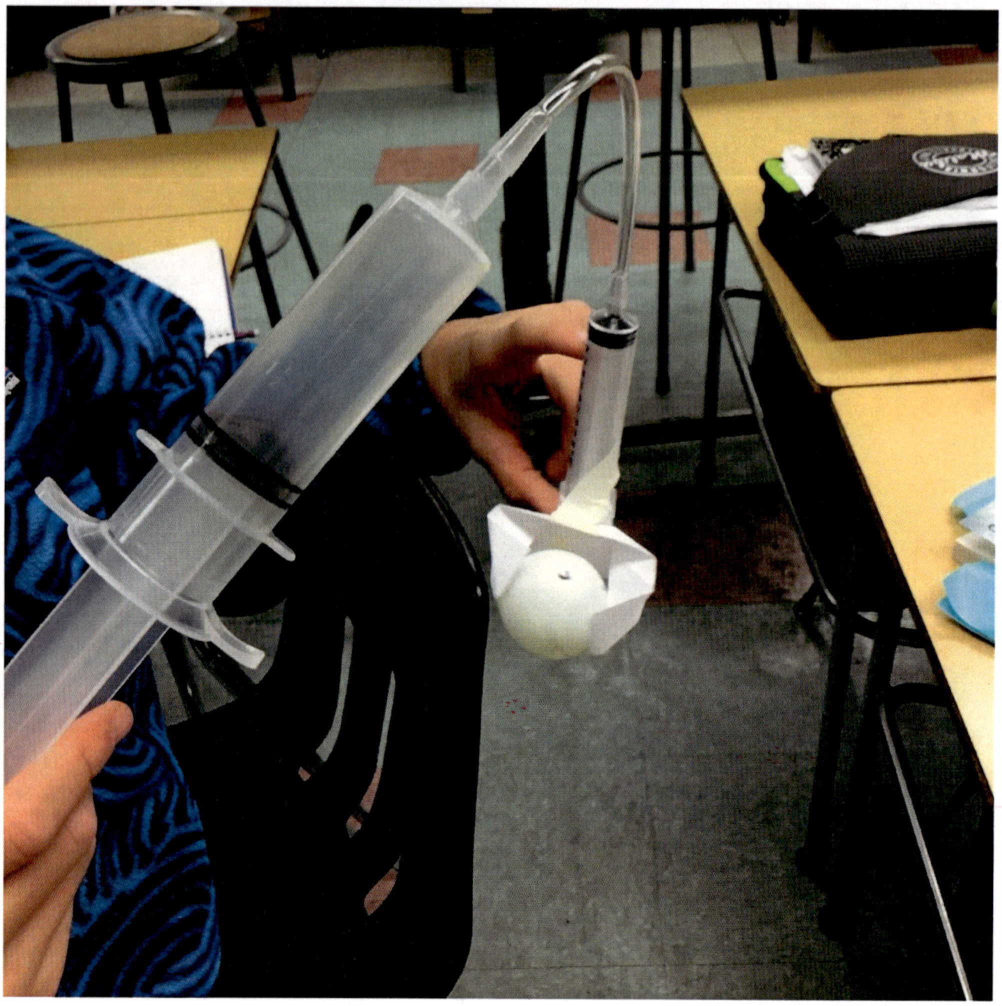

Or we could use scissors mechanisms made from tongue depressors.

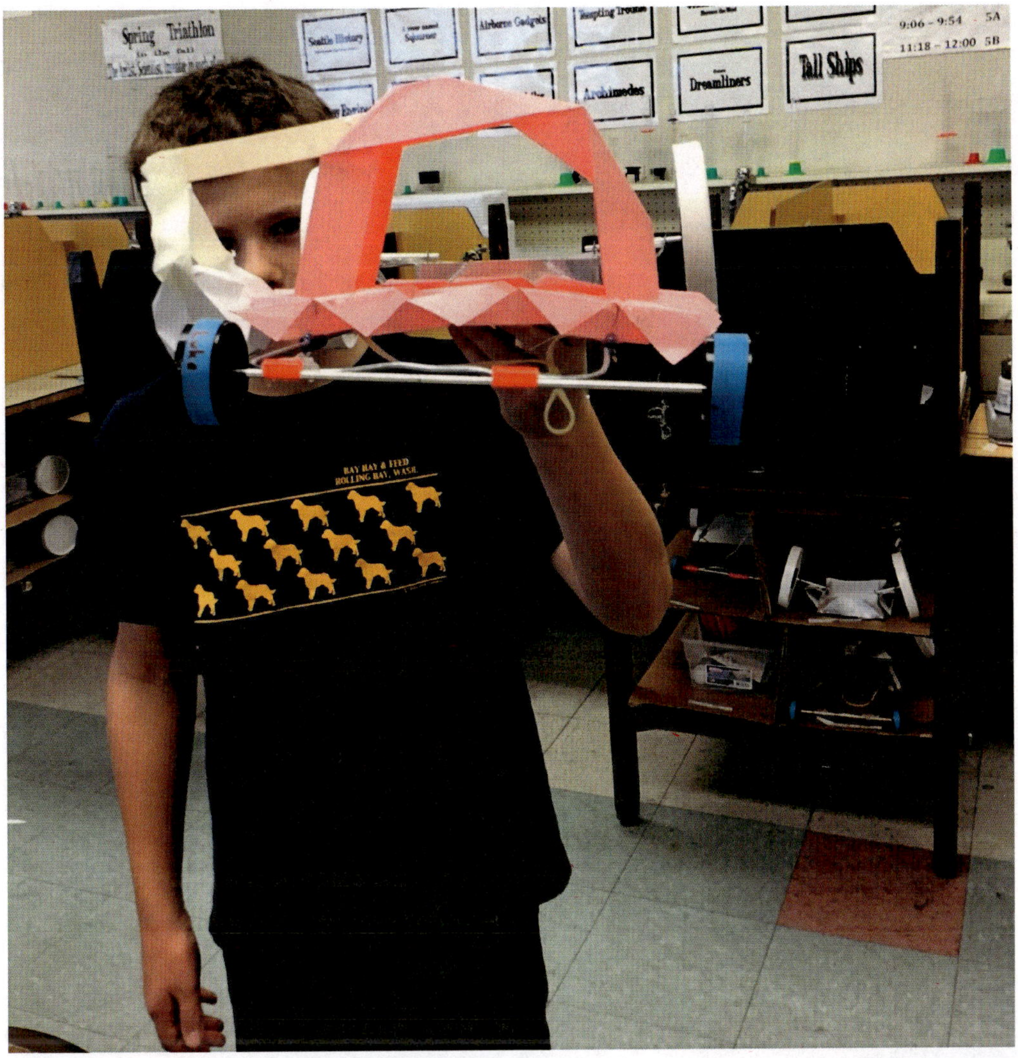
Spring Triathlon
Seattle History
Airborne Gadgets
Archimedes
Dreamliners
Tall Ships
9:06 - 9:54 5A
11:18 - 12:00 5B
BAY HAY & FEED
ROLLING BAY, WASH.

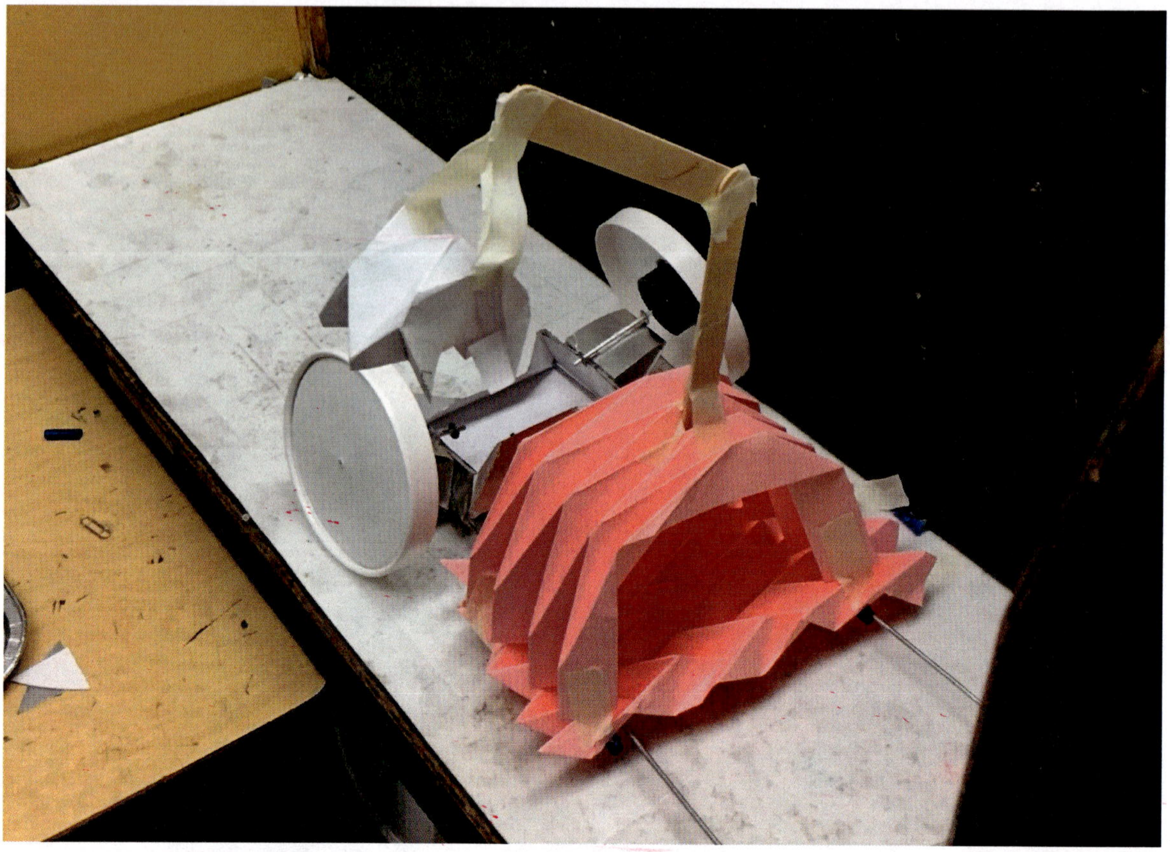

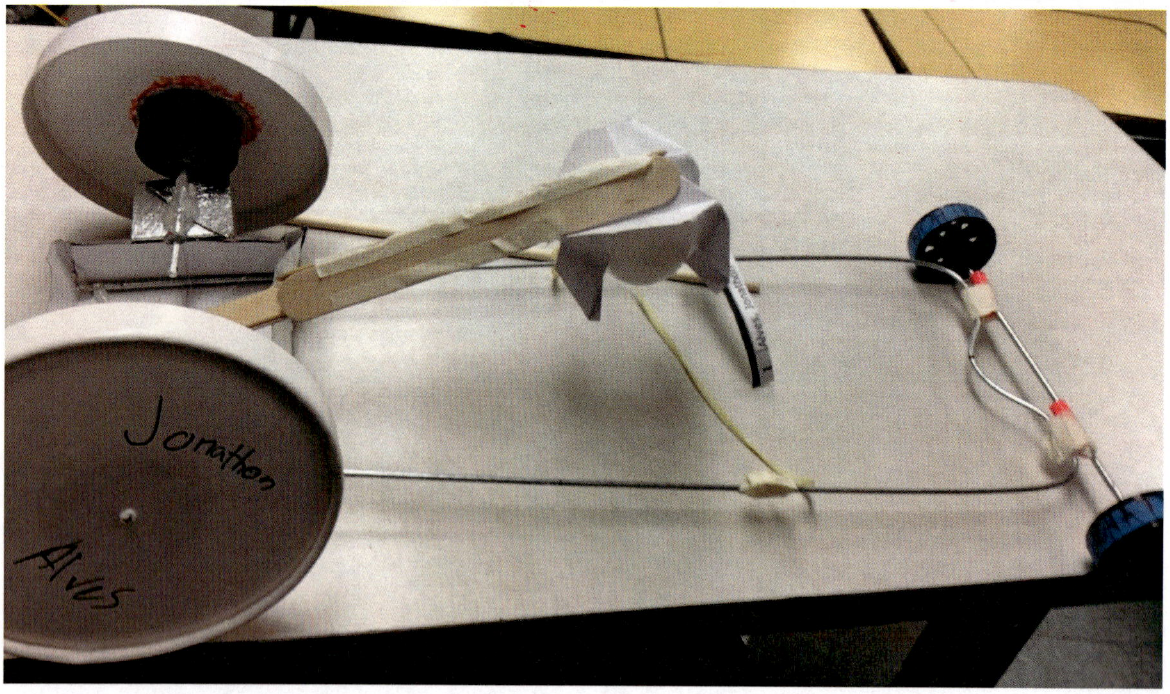
Jonathon
Alves

Tongue depressors can be used for scissors lifts.
What a good idea.
We are working on deployment. How to have compact things expand in space.
Scissors lifts would be good.
We could connect them to origami structures.

But, how to make scissors lifts?
We need a lot of holes in tongue depressors.
A drill press would work.
Our drill press has coke bottle “goggles” of its own to keep fingers safe.

Scissors lifts take shape.

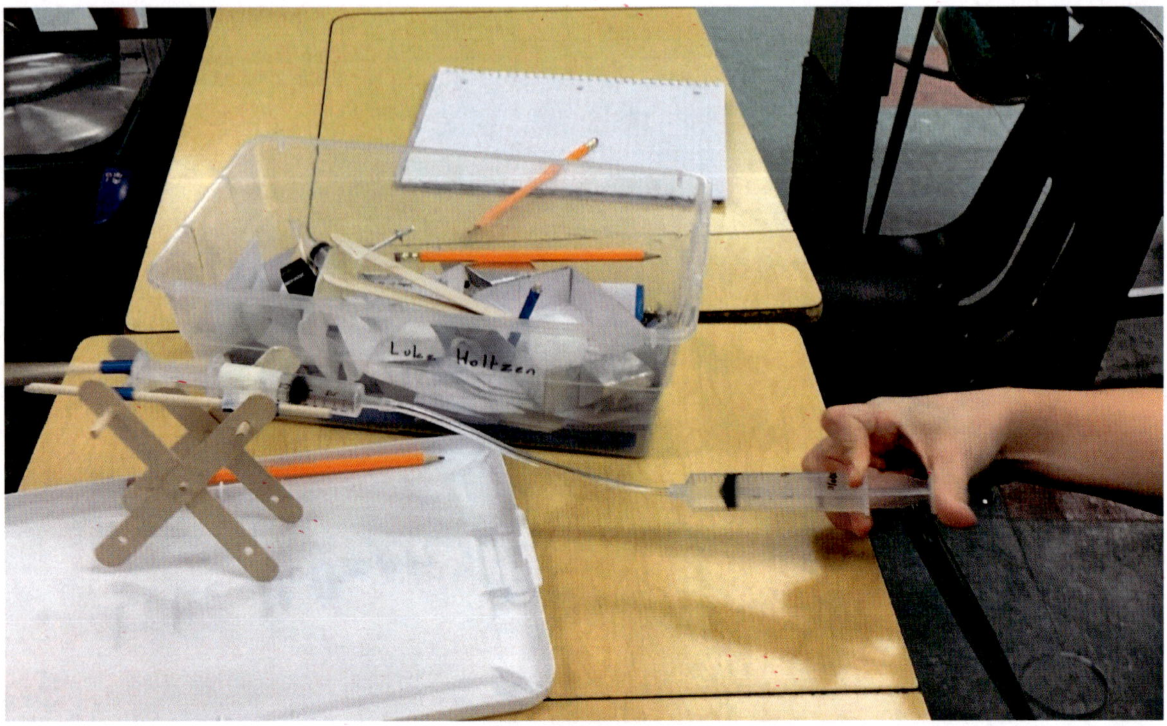

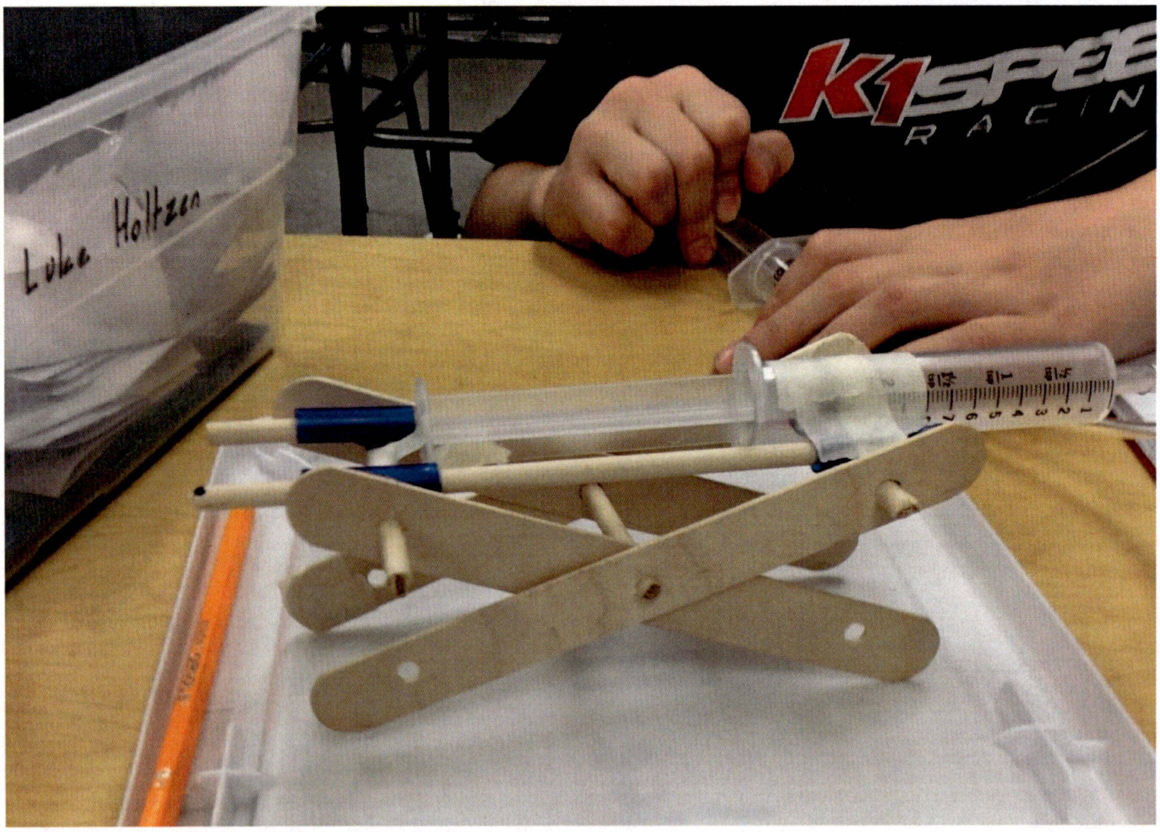

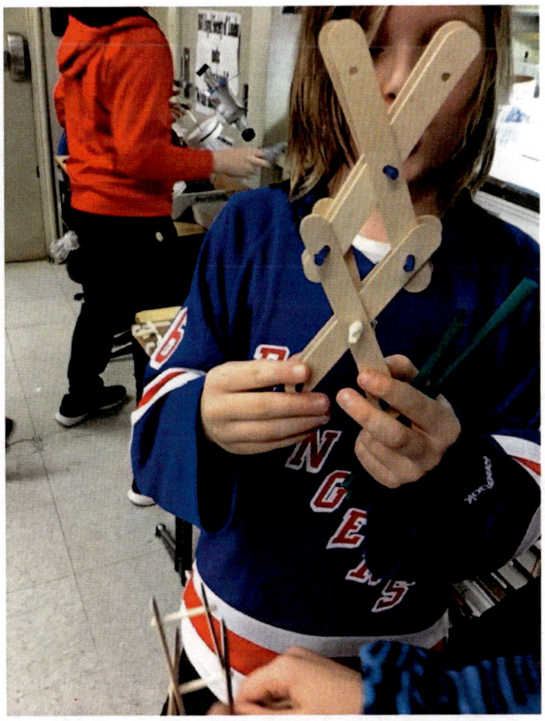

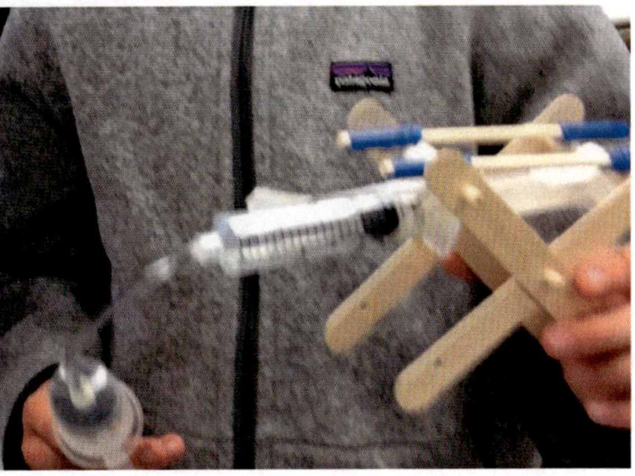

Tongue depressors come in handy for many inventions.

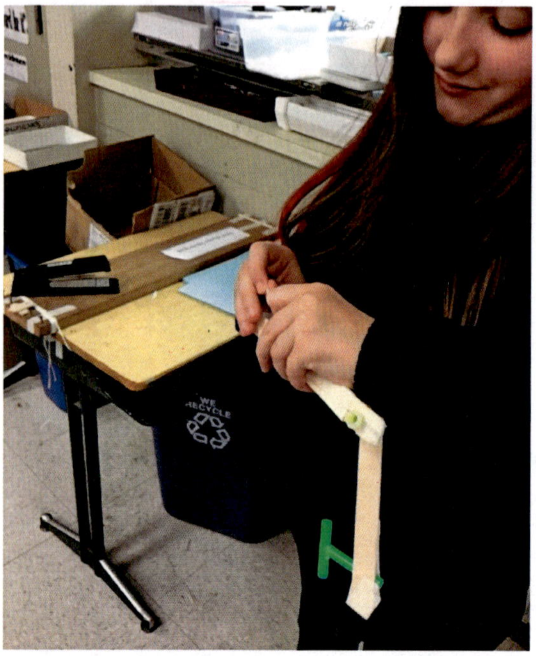

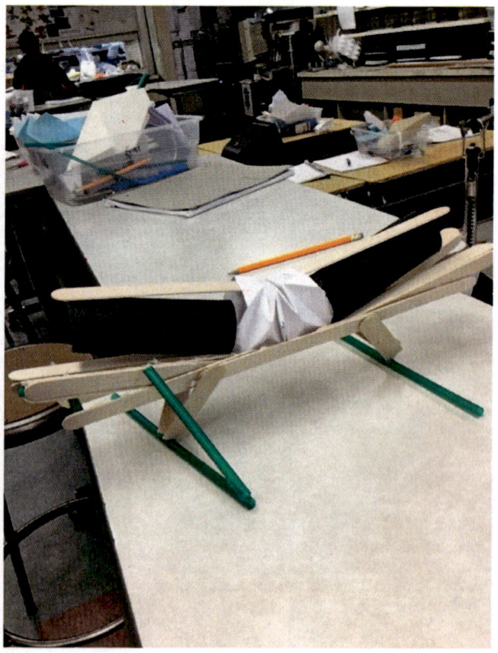

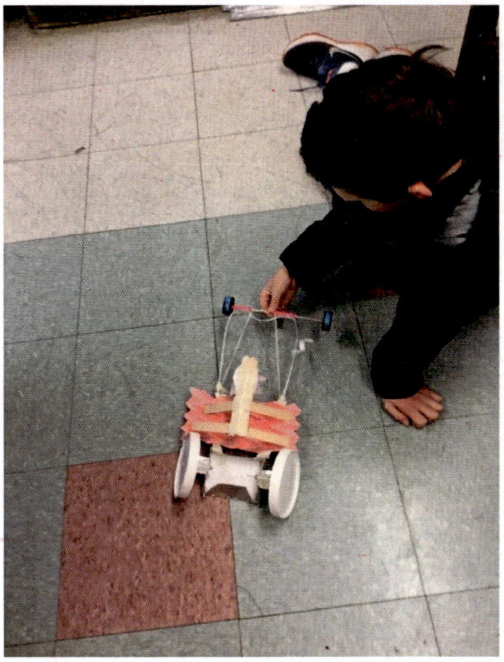

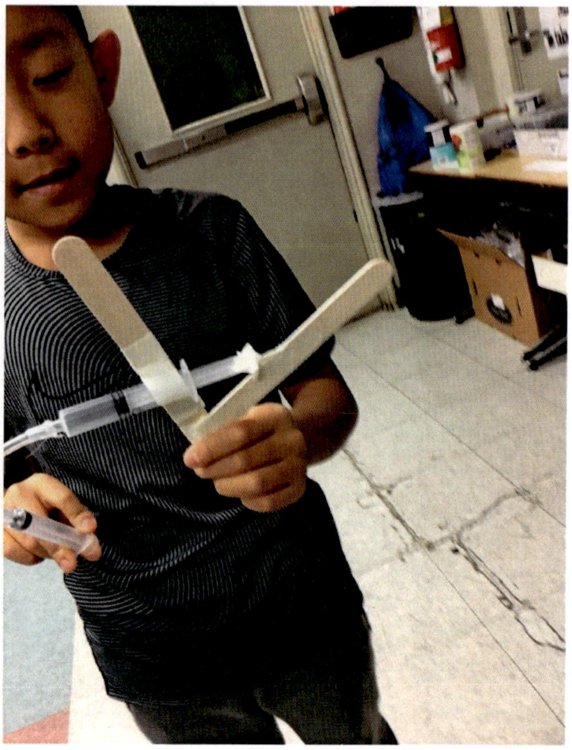

Some people got pretty good making the Miura fold.

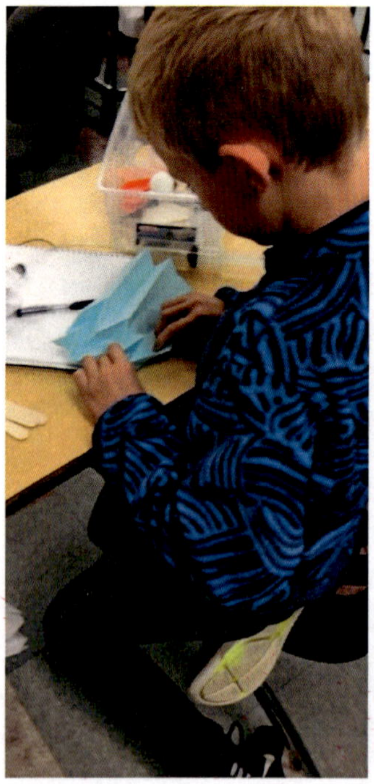

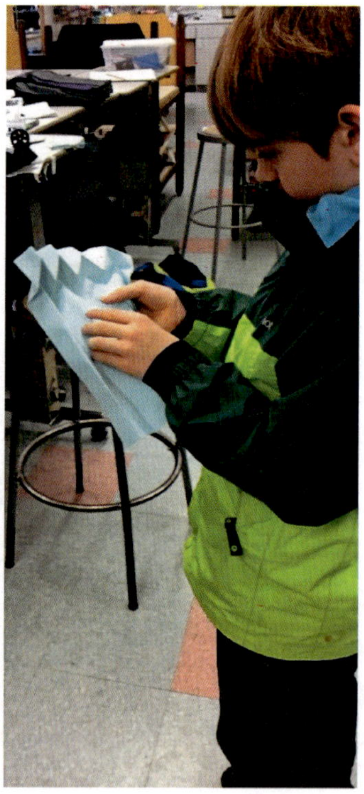

The challenge now is to combine all three ideas.
A moving rover, a Miura fold and pneumatic deployment.
Ideas flow.
Contraptions are amazing!

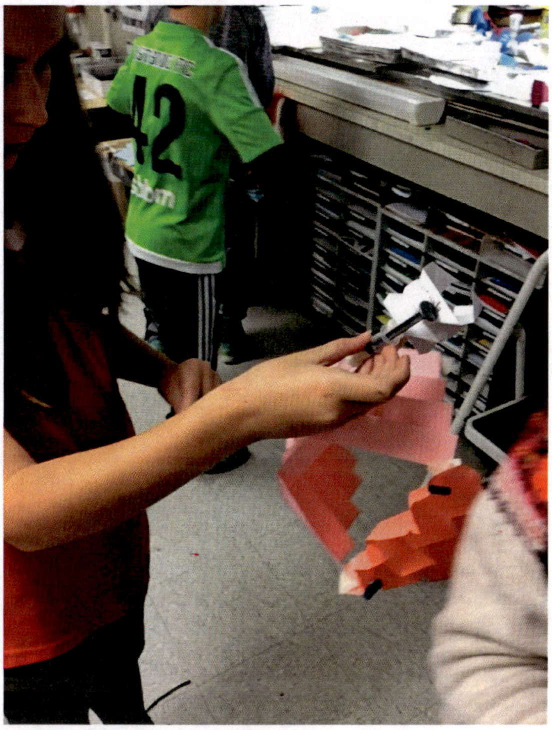

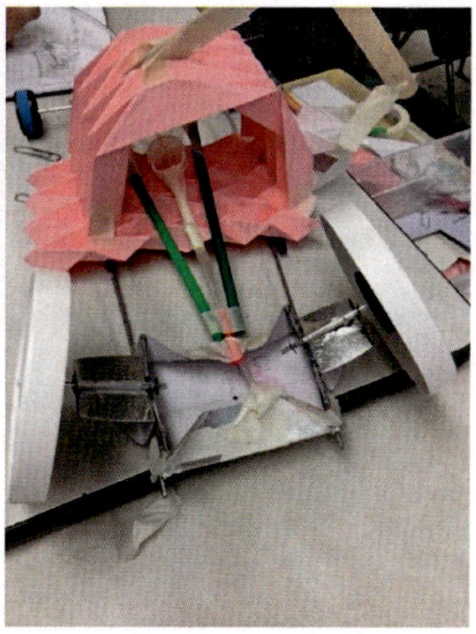

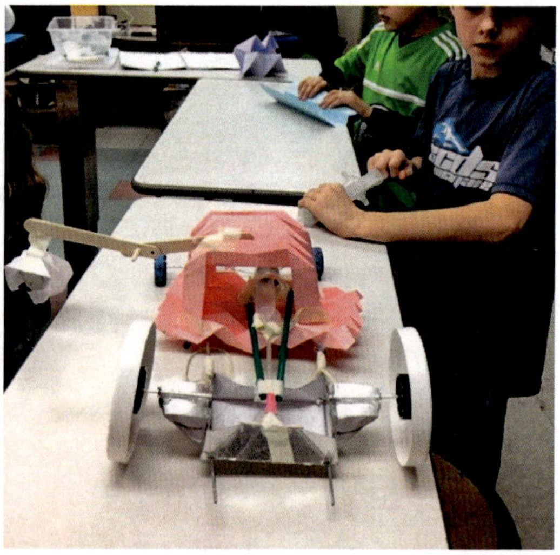

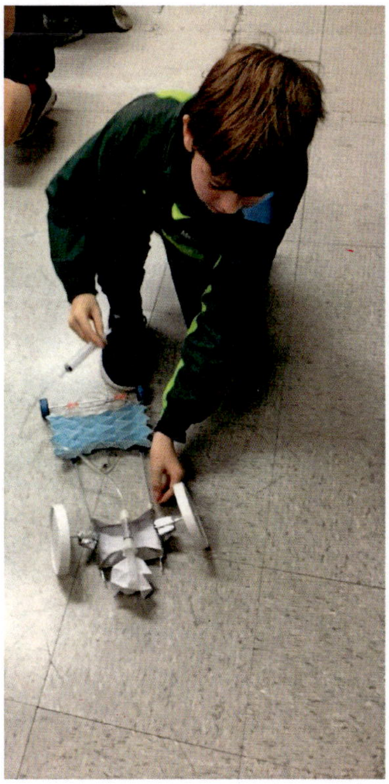

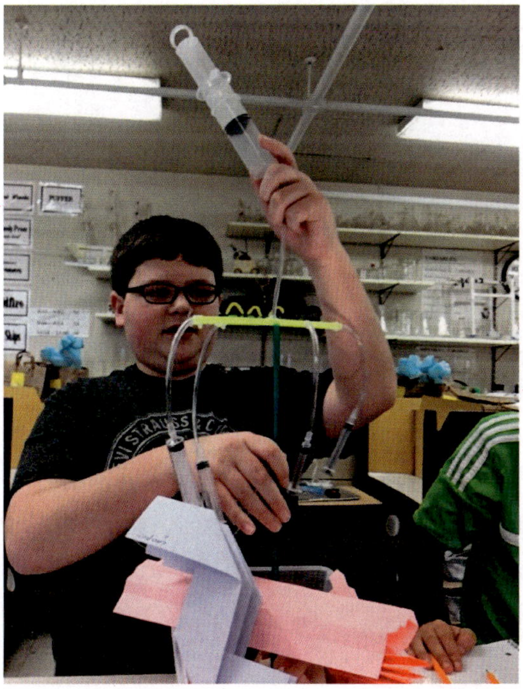
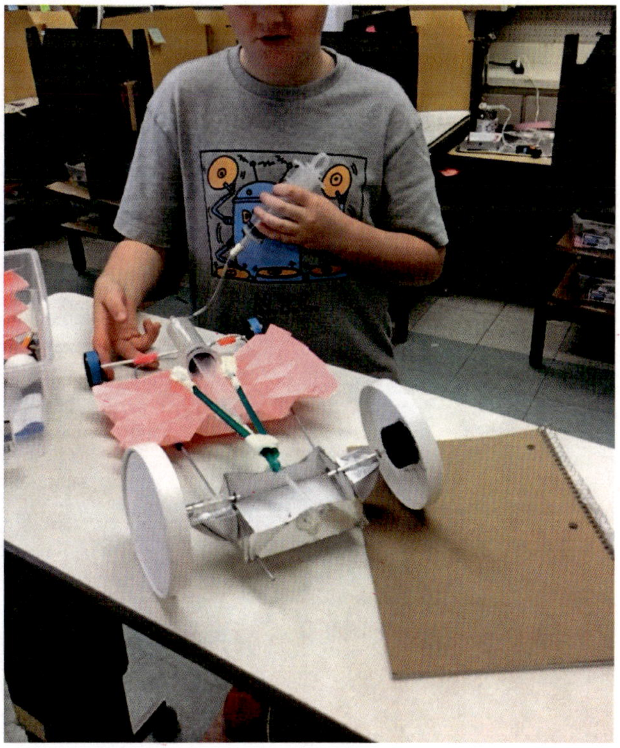

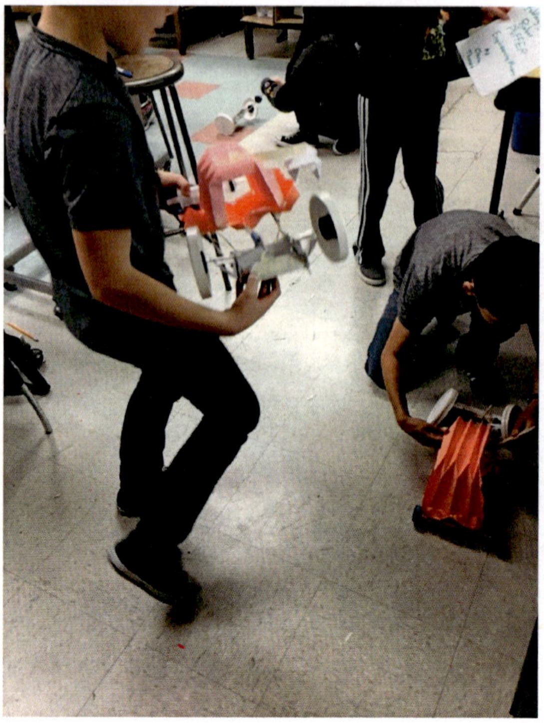

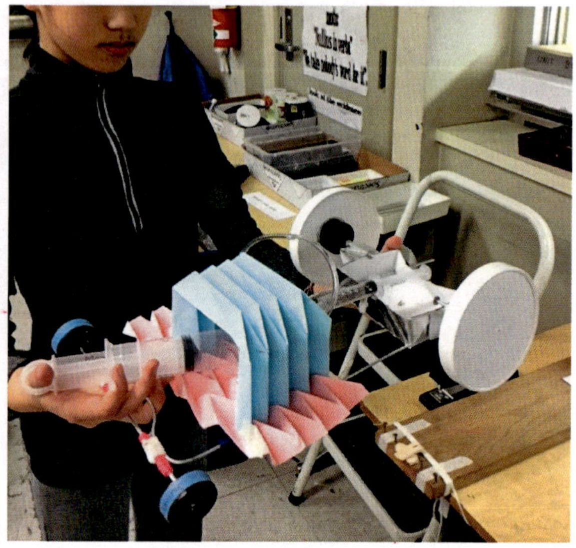

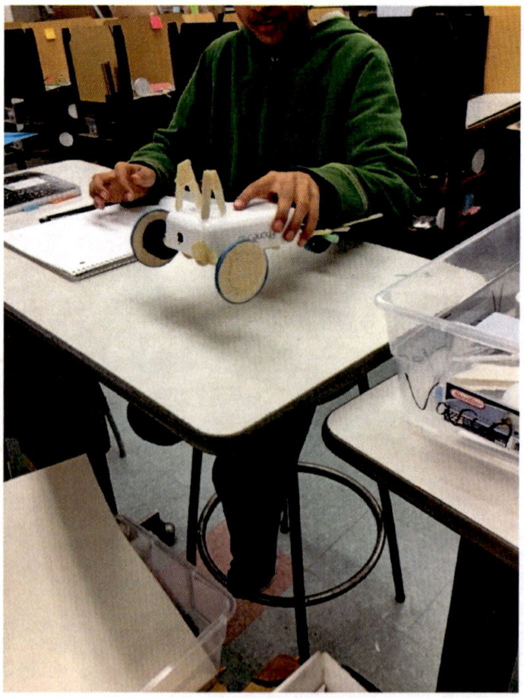

The gym event is next week! Time for "Rig Check" day! We need to see that all the things we have worked on for five weeks are still working. Ready. Set. Go!!

And then we go back to work. Fixing. Testing.

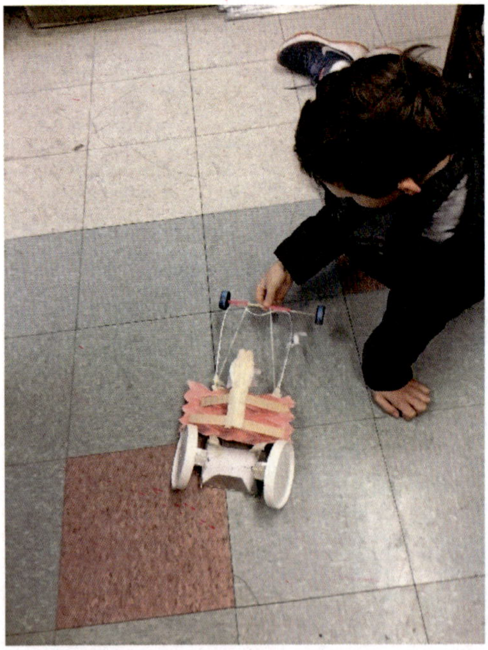

And origami. We need more Miura-ori platforms.

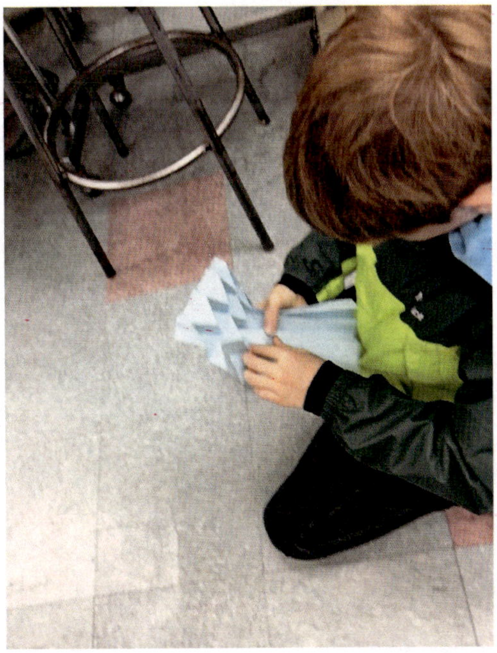

We help each other. When someone figures it out, they teach others.

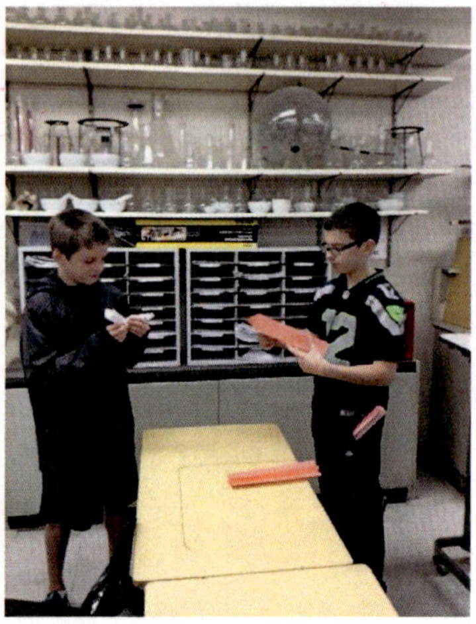

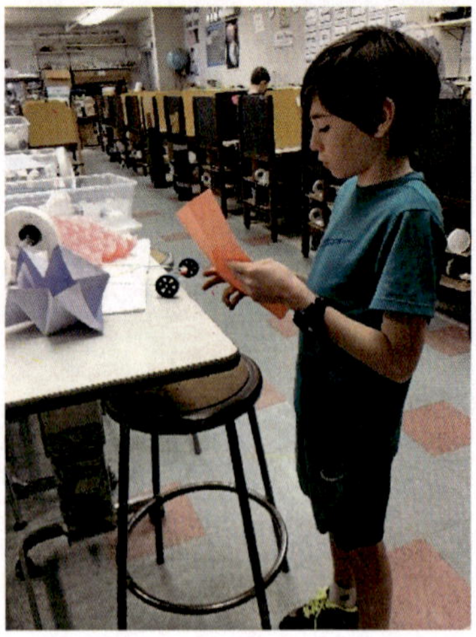

And more and more clever new inventions.

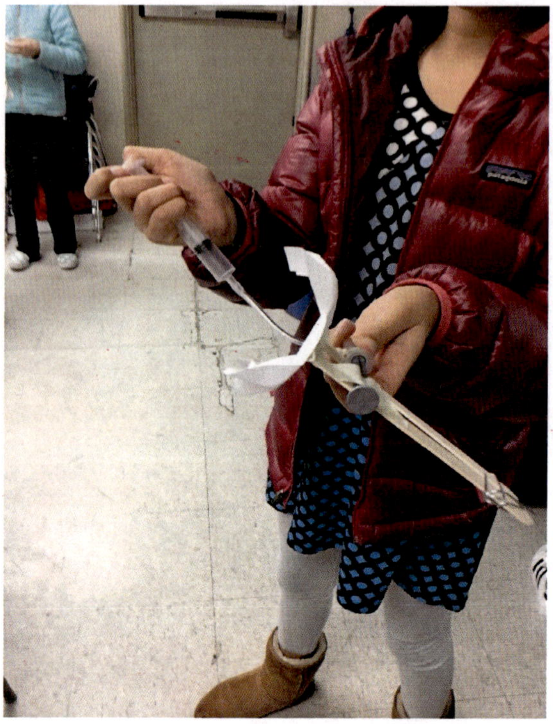
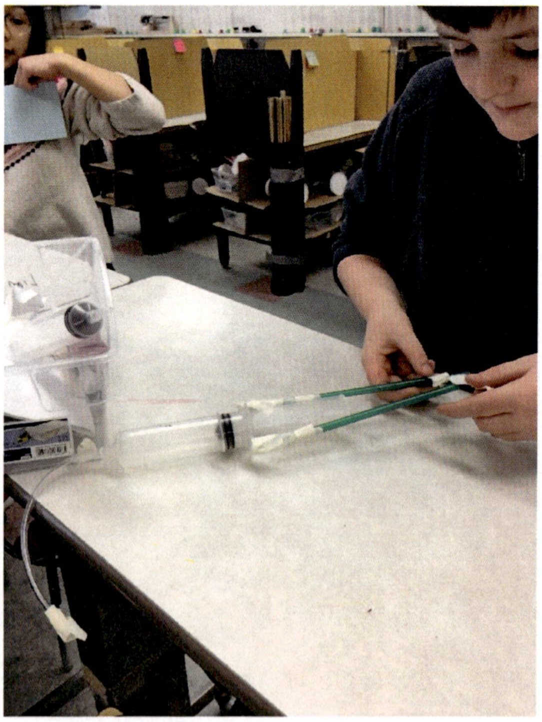
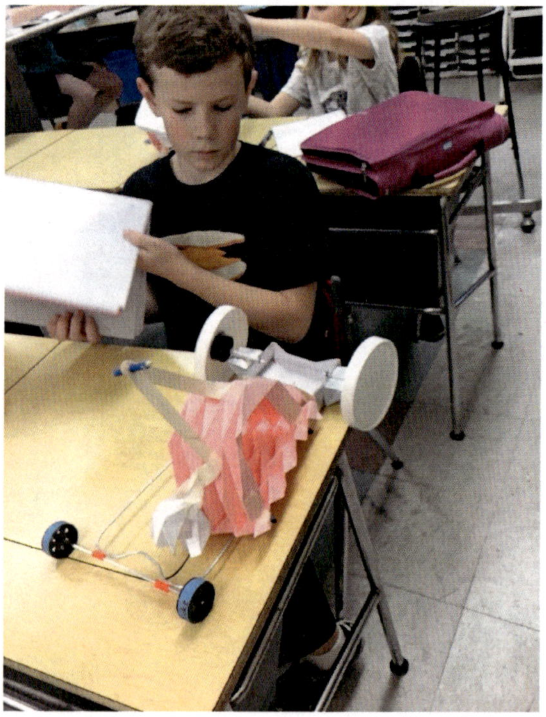

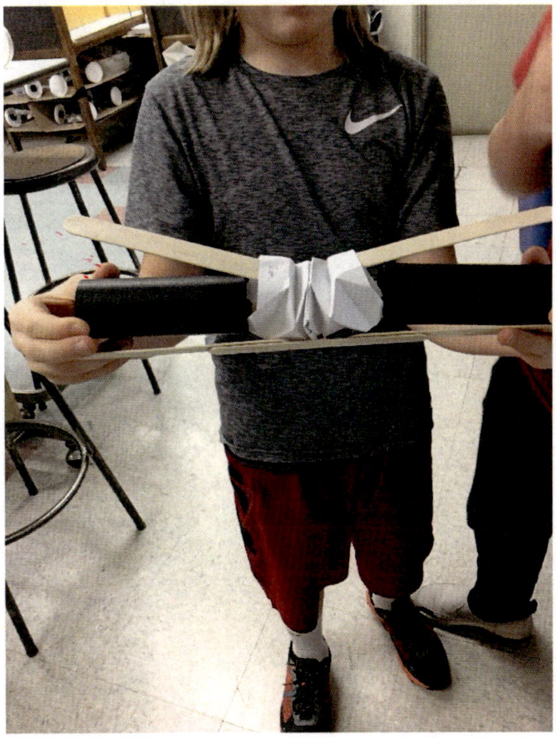

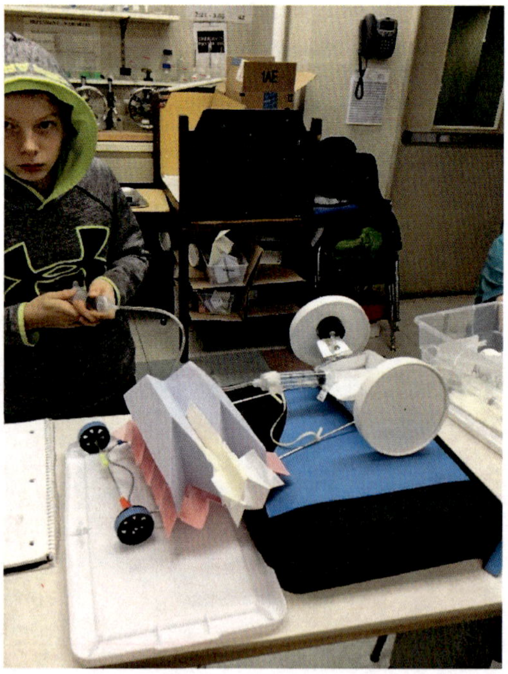
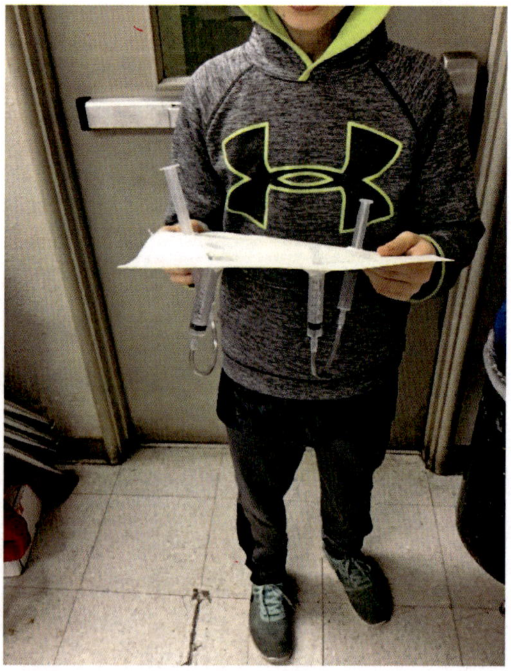

The Public Celebration

Time for a celebration in the gym.
We want to show our inventions to everyone.
"Puffer" makes us think of "Puff the Magic Dragon."
We can sing about our inventions.
We can make Miura-ori huts big enough for us.
The gym looks great.

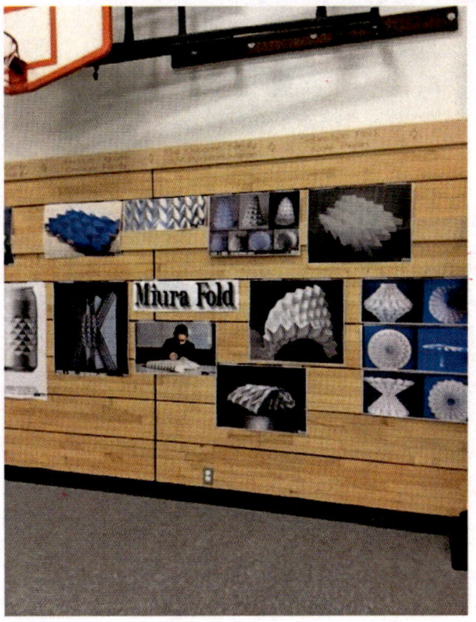

Tyvek makes a nice field to show our stuff on.

But we have to check to see if the rigs are ready

"Puff, The Pop-up Rover, Lived by the sea,
In A Miura-ori Hut, Big enough for me!"

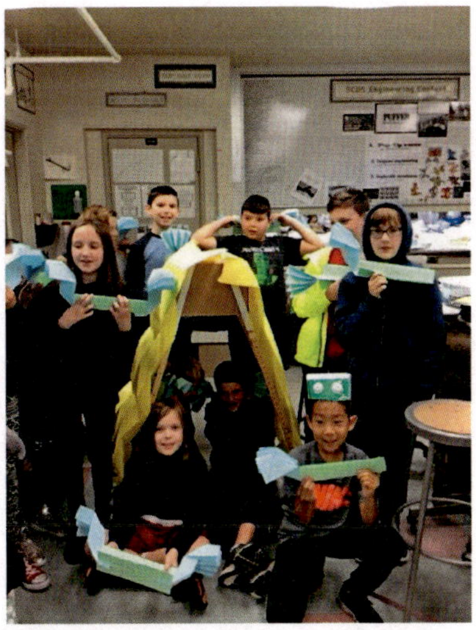

Yummy-Tummy Puffer

wonderful Pneumatics

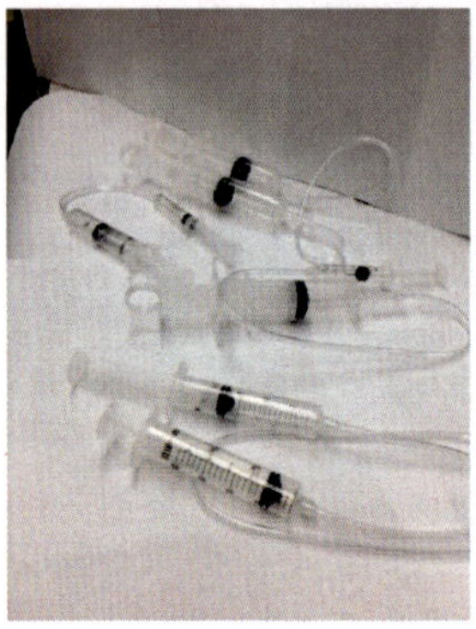

Learning the Miura fold and

how to engineer with it.

Celebrating Puffer
Pop-Up Flat-Folding Exploration Robot

Pop-Up Flat Folding Explorer Robot (PUFFER) was inspired by an origami design. This little robot could be used as a scout for larger rovers, going places that would be risky or hard to reach. Image Credit: NASA/JPL-Caltech.

and
The Miura-Ori Fold

About the Author

Meredith Olson Ph.D.

Dr. Meredith Olson, known affectionately as Doc "O" to her students, has taught elementary, middle school and high school math and science in Seattle for nearly 60 years. Her primary goal is in improvement of pre-college engineering education. By going to lab to work on contraptions every day, her students come to understand properties of the mechanical world.

"It has been a long and interesting trip. Studying some metallurgy in grad school. Evening classes. After a full day of high school teaching. Consulting for JPL as the Mars Pathfinder Educator. Weekends. Working in the summer with UNESCO in Zimbabwe, Kenya, and Uganda. Teaching dozens of weekend and week-long summer teacher workshops in South Carolina and Montana. Being a consultant and curriculum designer for Health and Physiology education in Washington, Oregon, Idaho, Montana, and Alaska. Being a summer adjunct University instructor for more than 20 years in Seattle, Idaho and Montana. Teaching teachers. Teaching students every day, every year for 59 years. Observing how learning happens. Becoming aware when real learning isn't happening. When it is just "show." When it is just teacher–pleasing to get a grade. To get a credit. To get a university degree."

See Dr. Olson's open letter outlining her philosophy of lesson design, available on the JPL website - Exploring Preface pp 11-13 http://mars.jpl.nasa.gov/education/modules/GS/GS07-19_preface.pdf

Dr. Olson believes that children must construct their own understanding from active design and assemblage of contraptions. By testing, failing, remodeling, and trying again, we come to see the structure when we look. By carefully examining materials we have, we may perceive how to use them in new and unexpected ways. Children begin to understand the engineering process. Besides, it is fun!

Made in United States
North Haven, CT
05 November 2026

21760279R00053